The Code of Creation With Guru Nanak and Albert Einstein

The Code of Creation With Guru Nanak and Albert Einstein

Two Supramental Visionaries

Book of One

Volume IV

Amar Kapoor M.D.

Copyright © 2020 by Amar Kapoor M.D.

Library of Congress Control Number:		2020901461
ISBN:	Hardcover	978-1-7960-8463-4
	Softcover	978-1-7960-8464-1
	eBook	978-1-7960-8467-2

All rights reserved. No part of this book may be reproduced or transmitted in any form or by any means, electronic or mechanical, including photocopying, recording, or by any information storage and retrieval system, without permission in writing from the copyright owner.

Any people depicted in stock imagery provided by Getty Images are models, and such images are being used for illustrative purposes only.
Certain stock imagery © Getty Images.

Print information available on the last page.

Rev. date: 06/11/2020

To order additional copies of this book, contact:
Xlibris
1-888-795-4274
www.Xlibris.com
Orders@Xlibris.com

CONTENTS

Prologue xi
Chapter 1   Portrait of Guru Nanak 1
Chapter 2   Portrait of Albert Einstein 12
Chapter 3   Portal of Truth 23
Chapter 4   God 28
Chapter 5   The Unified Field; Thought Experiments and Divine
            Singularity 41
Chapter 6   The Code of Creation 50
Chapter 7   SUPRAMENTAL Beings and Inner Technology 62
Chapter 8   The Royal Path 75

Epilogue 83
Bibliography 87
Glossary 91
Index 97

DEDICATION

This book is dedicated to:

The Generation One,

The Stewards of our Planet,

Humanity and Oneness!

and

Rinder, the love of my life!

DECODING TRUTH MATTERS!

TWO SUPRAMENTAL VISIONARIES:

One, a Nobel-prized physicist,

And the other

The Avatar of the Mystical world,

The Guru of Oneness and Truth

Are So Different

And Yet So

ONE,

Saturated in the essence of

ONENESS

Authenticated by the evidentiary

Existence

Of

The Unified Field of Everything and Consciousness

Are the ingredients of:

The Code of Creation.

PROLOGUE

MIND IS A magnificent and beautiful gift to humanity, of matchless resourcefulness. Mind has unlimited capacity of connecting with a source and creating one's own reality. It is a faculty of wonderment and awe. It can become expansive and all-inclusive. The mind has unlimited resources and an unbelievable capacity for manipulation, contemplation, and reflection. Unfortunately, the beautiful mind has been hijacked to the circle of make-belief and the house of illusive world. It has been contaminated by the descriptions of religions, the deception of dogmas, the deception of politics, and the deception by the hierarchy. It is beholden to the constraints of Dharma and Karma. The phenomenal god has been corrupted to the nth degree and there are millions of personal gods, each for a different reason or purpose. Such is the irreverence of god in the tomb of *Maya* and materialism, of caste and racism, of humanity fighting humanity – the endless wars, of world divided and fractured and fragmented. Such is the plight of humanity and our planet in crisis. At the end of the tunnel, there is a spark of hope with a beautiful mind seeking oneness and truth and purity of purpose.

Two supra-mental visionaries, namely Guru Nanak and Albert Einstein are a gift to mankind. One can easily harness the beauty and magnificence of the mind with their deep insight, thoughtfulness, and their purpose of life.

To enter the portal of truth, to seek the essence of One, the mind needs clarity, thoughtfulness, and a magnanimous vision of One, the mind of One. These two beautiful minds, Guru Nanak and Albert Einstein, gave us the guidance to unfold truth and oneness. This is the unfolding of two creative forces by two supramental visionaries. They gave us the transformative and royal path for transcending and connecting with the grand unified theory and the master-key of creation. Both exceptional beings, Nanak and Einstein, on different paths seeking truth and the essence of truth present to us radical possibilities to be explored on the cosmic platform of One. This indeed is the royal path to the awakened state with pure awareness and consciousness. The two creative forces when harnessed can become the master key that unlocks the greatest mysteries of the universe, for everything is connected in the Unified field and the core of the infinite One.

Come join me to explore, expand and enlighten on Nanak's code that has remained encrypted, and Einstein's gift to humanity that can give the vortex of power from the energy of existence. You can supercharge your whole being when you are anchored in One, and connected with the unified field of One.

Let us explore the realms of transcendence with mindful meditation of Nanak, which he called "Naam Simran".

You can reverberate and resound the holy essence of the true name. Einstein gave us a code of the Unified field, called the general theory of relativity in a very simple and elegant equation:

$$E=mc^2$$

where E stands for energy, m for mass, and c^2 for speed of light squared; this is a cosmological constant. We do not know what is energy or what is a mass and the possible cosmological constant, the speed of light. The quantum physicists may have differences but it is a done deal with the measurement of gravitational waves in 2016, for the unification of Albert Einstein's dream was fulfilled. The dream was fulfilled with the measurement of gravitational waves for the unification of unified field theory. Unified field theory was coined by Albert Einstein who managed to unify his general theory of relativity with electromagnetism. In this 100 year old unification journey there were many, many mathematical models, many hundreds of physicists with denials and disagreements from space-time continuum to gravitational curvature to dimensionless particles or discrete points of nothingness. But we have arrived to an understanding of hyper space continuum and the Unified field.

It seems everything from the electrons to the subatomic particles is in a state of vibration. Even the celestial bodies in distant galaxies are in a state of vibration. The name of One is encoded in vibrational frequencies with layers and layers of sound currents. We also know sound is the creative force. The most creative sound force was Big Bang. Sound precedes light. There are few and rare ones who are aware of Nanak's code, which is enshrined on the pages of Sri Guru Granth Sahib, the holy book of 31 million practicing Sikhs. This code is 550 years old. Then there was Moses code, which was, "Ehyeh Asher Ehyeh", which means I am that I am. This is probably as old as 3500 years. And yet, the code was only recently revealed to the general public; only few Rabbis and privileged were privy to the code.

Prior to this code there was Brahama's code, "Aham Bharmasmi", which means I am that or I am the universe. And this code is 5,000 years old. And yet, few people believe in it or practice it. Then we have the Bible code, uncoded in 1994. There were visceral attacks or beliefs of non-believers and believers. But we have new technology. We will explore the depths of the code of creation. The code of creation uses the pole star of true name, the essence of existence, and the current fountainhead of knowledge.

Combining Einstein's coded Unified field and Nanak's Code of the core of One, we can access the code of creation. We will go into the details of what is the Code of creation and how we access it. Einstein famously said, "I want to know God's thoughts, the rest are details." We will delve into Einstein's thoughtfulness and harness the great treasury of the Vacuum Space that never empties, and with Nanak's insightfulness:

"EK OM KAR
SAT NAM"

This is the master key of creation. Please join me on this enlightening, empowering retreat of eternal quest;

the code of creation,
the font of absolute knowledge
the field of pure consciousness,
the region of pure knowledge,
the wellspring of quantum field
with zero-point energy of Vacuum space
whirling and vibrating
in the celestial vacuum energy.

This is indeed the universal heart of pure existence, infinite intelligence with the immensity of consciousness.

As time and space are nature's creed,
truth and reality sparkle on the twin seekers,
one on the mystic path
the other seeking the one with the tools of science. Both are unified
at the core of One! Look within
the pole star resonates
with light!
Sound and light in Quantum entanglement
reveal the secret
of quasars and mysteries within,
with the Avatars and Prophets
with thought particles when attended to collapse into thought waves
where patterns emerge
of codes and fields unthought of,
infinity upon infinity,
resonating the vibrational frequency of the eternal atoms
in all probability fields of thoughts and patterns embedded and
encrypted in the DNA of the eternal one!

We are the waves of the ocean of the unified field of energy, infinite intelligence, and consciousness.

The infinite Oneness is within us.

The essence of existence is beyond our thinking and is barely unfolding in the vast mystery! Science and spirituality are merging into a Unified field, the Ek Om Kar and the Unified field of Oneness.

The core of the being is a quantum soul, the source of the infinite.

One is all!

All is one.

From Big Bang to the vast existence, we are the expressions of the universe.

We are the one in the inner cosmic web
expression of a single field of energy, all particles, everything is interconnected
through an underlying existence of the Unified field and Infinite Oneness.

Decoding Truth Matters
Two Supramental Visionaries:
one, a Nobel-prized physicist
and the other
the Avatar of the Mystical world,
the Guru of Oneness and truth,
are so different
and yet so
ONE,
saturated in the essence of
EXISTENCE!
Herein are the laws of Nature and the Treasury of Consciousness that is the crux and the crucible of the Code of Creation!

PART I

TWO SUPRAMENTAL VISIONARIES

CHAPTER 1

Portrait of Guru Nanak

Preamble to Japji (as in Mool Mantar)

There is One God, the Eternal One.
Eternal Truth is the Name
the Creator of all things
Fearless is the One
without animosity
Timeless Being, the ultimate reality
unborn
self-existent
Obtained by Divine Grace

Mool Mantar

Ek Om Kar
Sat Naam
Karta Purakh
Nirbhau
Nirvair
Akal Murat
Ajuni
Saibhang
Gur Prasad———Guru Nanak

Introduction

GURU NANAK is the prophet of Oneness and truth, the last Avatar of the mystical universe, and one of the greatest saints. He had a very powerful message in his signature codified phrase which literally is the opening of every stanza in the holy book of the Sikhs, Sri Guru Granth Sahib. The mystic saint encoded the message in his prophetic phrase that needs to be unfolded for our present time. Guru Nanak was a great mystic and reached the level of a very advanced yogi, the highest echelon of a yogi. Nanak's devotion and divine love for Ek Om Kar is matchless. Nanak was the embodiment of Ek Om Kar. Nanak was not only a yogi of love and devotion, which is called "Prema Bhakti" (devotion by love), but he had achieved complete union and total alignment with the cosmos to become a body of light, sound and energy. He touched the higher regions of consciousness and total bliss. Nanak imparted us with a royal path of union with the most high, with his inner technology, which is "Surat-Shabad Yog".

Guru Nanak has a message which is going to transform humanity at crossroads of die-in-fear or live-in-bliss. This chapter is not about Guru Nanak's life history or his birth story, which have been covered in multiple books, (1-15.) Rather, it is about his divine transformation and enlightenment. His mystical philosophy and his spiritual exaltation with the codified word will be revealed for the greatest good of humanity. His mission was simple. He wanted to awaken divinity in all people, in all races, in every village on the planet with the incessant call and quest for the essential divinity of oneness. He believed in the service to humanity, sharing with humanity and caring for one and all.

House of Nanak

It happens that 2019 is the 550th birth anniversary of Guru Nanak. Nanak was born in 1469 AD in a small town called Talwandi, now in Pakistan. The town is now known as Nankana Sahib in honor of Nanak, which has become an important landmark for pilgrimage.

His father Mehta Kalu was of Hindu heritage. Mehta Kalu was an accountant. Mehta Kalu went to his religious brahmin pandit to inform him about the birth of Nanak. Pandit Hardayal charted the course of Nanak and created his horoscope. And it is said he wrote, "You are truly blessed, Kalu – a great and divine soul has taken birth in your house today. Your son is born to a greatness that comes to the world but rarely. His light will spread far and wide, bringing radiance to those dark times. Your son will be revered by both the Hindu and the Turk," (1-5.) This reminds us of the three wise magi who followed a star to find baby Jesus lying in a manger in Bethlehem in Judea. That marked the birth of Jesus, and the miracle of Virgin Mary. It is to be realized that during this time there were caste systems and bitter hostilities between Muslims and Hindus. Mehta Kalu's new neighbor was a Muslim couple by the name of Sayad Hassan. He was a grand scholar and had a library in his house. Sayad Hassan played a key role in Nanak's early childhood and personal education(4). Nanak was sent to the village school at age 7. At age 9, he composed 35 stanza poems using 35 letters of Gurumkhi (Punjabi language) script, which dealt with universal truth and spiritual meditation called "Patti Likhi" and was included in the holy scripture, Sri Guru Granth Sahib(1). His teacher Pandit Gopal was amazed with the spiritual content of the poem and admitted the child was more advanced than him. At age 9 years, Nanak quit school. At this time in his life he became very interested in going to the forest with the village holy men and ascetics to discuss sacred literature. Nanak's father was very concerned about the future of his son. He did send him to learn Sanskrit and Persian. He learned Persian from Maulana Qutab-ud-din who also introduced him to Quran, the holy book of Muslims.

Next in his life, at the time of puberty he attended a ceremony for the brahmins, a sacred thread ceremony called Janeu. The sacred thread was a symbol of superior caste and Pandit Hardayal was to perform the ceremony. After the holy rituals and the recitations, Pandit Hardayal was about to place the thread on Nanak's neck when Nanak held up his hand to stop and asked him about the significance of the thread. The Pandit declared that the thread protects one and distinguishes one as an upper class Hindu and by wearing the thread you become a pure Hindu. It was a sacrilege to refuse the sacred thread. To this Nanak replied, "Does a person become pure merely by wearing a thread?" The Pandit felt very insulted and left the thread on the floor. Nanak composed a message for the Pandit which is again written in the holy text Adigranth(1);

> "From the cotton of compassion
> spin the thread of contentment
> let continence be the knot with twists of truth.
> This is a sacred thread of the soul.
> O Pandit, if there was one such thread
> then put it on me,
> a thread so made will not break
> nor soil, nor will it be burnt or lost.
> Blessed is a person, O Nanak, who wears such a
> thread around the neck."

This is on page 471 in Asa-di-var.

Nanak continued his discourse with the village holy men, *sadhus*, and ascetics in the forest. Nanak's father lost all hope in his son's future development. Nanak had a childhood friend by the name of Mardana. Mardana was from a Muslim family of lowly *marasis* or singers. Nanak loved his singing and formed friendship with Mardana. They continued to grow and mingle with each other(12).

Nanak was aloof most of the time and wandered in the forest. Nanak's father tried to occupy his son in activities like cultivating land or herding cattle and shopkeeping or dealing with horses. He refused to take up any of these professions. Nanak grew up as a precocious child. He squandered his father's money by feeding the poor and the hungry. Nanak's father was thoroughly frustrated with his son and had given up on him. He let him spend time in the forest with the holy men.

He used to spend a lot of time at Mardana's house. He was fond 0f singing with Mardana. He had another childhood friend, Bala, who would go to the forest and buy food for the starving *sadhus*. This angered his father who was continuously losing his money on mendicants. At this juncture, Nanak became more withdrawn, would go off to solitary walks. He stopped all of his outside communication with his friends and community. His family and close friends began to think he was possessed by an evil spirit. So the family physician was called to examine Nanak. When the physician checked the pulse, Nanak told him that he was afflicted not with the pain of the body, but with the pain of the inner being, the soul. Nanak continued to spend long hours in inner introspection. *Nirbanis*, a sect of holy men who live with nature, do not take anything of material nature, do not even wear clothes and live on berries and fruits, fasting if no food was in sight. Nanak would feed them and said that was a true bargain. Nanak had an older sister by the name of Nanaki who took keen interest in his well-being and suggested that he should married, and even found a match for him. Nanak was reluctant to get married but conceded to his sister's request and got married at age 17 to Sulakhni. Nanak shared his feelings and would commune with her. He tried to tell her that he was trying to lead would be revealed and lead to salvation. However, his mother Tripta was very concerned that he did not have the right education or a job since she was deeply concerned about their marriage. Nanak's sister took Nanak to Sultanpur, another small city in Pakistan, this time to find a job. She found him a job as a storekeeper with the Nawab of Sultanpur, who was the governor. It was Nanak's job to measure and weigh all the grains that were to be distributed and put into sacks, as the store transactions. He would reach 13 or *tera* in Punjabi. He kept on counting, "*Tera, tera, tera, tera,*" over and over again. That was really addressing God, *tera* translates to, "Thine, thine, thine thine. All is Thine."

Nanak started composing hymns in praise of God. At this juncture, his friend Mardana joined him again and played *Rabab*, a stringed instrument. Nanak started having small gatherings at his house and they became bigger. People started to come from faraway places to listen to Nanak's discourses and singing God's praises. Professionally,

Nanak was doing well and had a reputation for meticulousness and transparency in the conduct of all his transactions and record keeping. It is to be noted that Nanak developed a phenomenal record keeping system. In 1494, Nanak was blessed with a son. He named him Sri Chand. In two years, he had another son named Lakhmi Das.

Nanak immersed in more deep meditation, radiating an aura of profound peace. A significantly spiritual event happened to Nanak. Nanak had gone to the river Bein for his early morning bath. He had folded his clothes and kept them on the riverside. He did not return home for breakfast. So Mardana went down the river to search for him and could not find him. Then the whole neighborhood and the village started looking for Nanak. For three days people could not find him and they thought he probably had drowned. At this time, the governor's soldiers looked for him everywhere and one of the soldiers found him in the forest sitting cross-legged in deep meditation. He was in *samadhi* – deep meditation. When they questioned Nanak what had happened, he remained silent. The *Puratan Janamsakhi*, which is the record keeping of Nanak's life history described the incident in great detail. And when Nanak came back to speak, he would utter, "There is no Hindu, there is no Mussalman." The citizens of Sultanpur thought Nanak had lost his mind. And according to Macauliffe, (14), Nanak received his proclamation and it had deep significance. The *Mool Mantar*, the opening preamble of the sacred prayer *Japji* was channelled to Nanak. Nanak spoke about his strange experience and the inner voice kept on telling him to deliver the message and teachings to every village and town." Nanak uttered the first word, "There is one god, there is no Hindu, there is no Mussalman."

The *Qazi* or the Muslim religious authority was very perturbed and reported the matter to the governor and felt this was a sacrilege of the worst kind that there is no Muslim. Nanak had to address the *Qazi* and go to the Muslim congregation for a Friday prayer in the mosque. When all the worshippers knelt in prayer, Nanak remained standing. The *Qazi* was upset and told the governor, "My lord, this man is a liar. He said that for him all religions are equal, but he did not join us in the prayer." The governor asked Nanak to give his explanation. Nanak asked the Q*azi*, "You are a man of God in the house of God and you led a congregation of a thousand men in prayer. Tell me the truth, what was in your mind?" The Q*azi* replied, "I was thinking of my mare. My mare gave birth to a foal and I was worried that the foal would fall into a well and I would lose him." "So tell me, learned one," Nanak asked, "does prayer consist only in kneeling and bowing down and reciting a few words?" "No," said the Q*azi*. "True prayer comes from a focused mind. So that when you praise God, you think only of him and of nothing else. You are right, Nanak. While my body was bowing to God, my mind was full of other things." "And do you become a Mussalman by merely reciting the N*amaz* five times a day?" The Q*azi* pondered and had to give a truthful answer, "No. We must be firm in our faith, our heart must be cleansed, and we must not have any greed or pride. We must accept the will of God. We must be unselfish and kind to all. Only then can we call ourselves true Mussalmans." The Q*azi* understood the depth and import of Nanak's word. Nanak was given the message in his deep meditation to spread *Naam*, the true essence of God, the true name of God, that there is only one God! Nanak had to complete and spread the mission of one God. So in 1496, Nanak with Mardana, his constant companion, set out on the missionary journey. Nanak was on a special mission to take the message of truth as was given to him during his deep spiritual awakening. He spent 23 years of his life on the road travelling to hundreds of villages, towns, cities, different countries, including Tibet, Ceylon, Saudi Arabia, Iraq, Iran, and some places far away. He travelled on foot, boats, dinghies, or horse-driven carts. Nanak was accompanied in all his travels by Mardana, who played the stringed instrument, the *Rabab*.

It is to be noted that he wore monk's loose "choga", worn by Muslim dervishes, of saffron color of Hindu ascetics and white cloth belt around his waist, worn by fakirs. On his head he wore the cap partially covered by a turban similar to Sufi, and on his feet wooden sandals as favored by religious men. His outer continence and dress emphasized the message of oneness, of there is no Hindu, there is no Mussalman(12).

Many scholars including Bhai Gurdas have written extensively on his travels or what is called *Udasi*. There are different versions of his extended forays. The first marathon journey took 12 years, which commenced from Sultanpur to Lahore and then back to an area called Amritsar wherein lies the Golden Temple now. Nanak and Mardana rested under a tree near a pond in Amritsar. Nanak sang a hymn in praise of God who had created this beautiful world. Guru Arjun Dev selected this area and the tree under which Nanak stood and sang a hymn, to build the Golden Temple or *Harmandir Sahib*. And the tree under which Nanak rested still stands in the compound of the temple. Nanak went back to *Talwandi*, to his village, to meet his parents and friends. The chieftain of the village, Rai Bular, was very sick and came to meet Nanak. Rai Bular was totally moved when he met Nanak and asked him, "What is your path and perception of truth?" To this Nanak responded with considerable deep thought; "There are four stages on the path to salvation. The first is D*haramkhand* (discipline). Second one, G*yankhand* (knowledge). Next one, K*aramkhand* (action). These lead to the fourth stage of blissful merger in God. The fourth stage is *Sachkhand*, which is the region of Truth and the residence of the Lord Master. Rai Bular became a disciple and accepted him as his guru. There are anecdotal biographies and writings by Gurdas and there are contradictions in his life history writings. These narrations with myths and miracles and imaginary legends were written by Nanak's followers lacking any historical exactness, so I will not venture into these historical details, but I'll emphasize the spiritual legacy of Nanak and of the message of his teachings. It is to be emphasized that his travels were extensive. His message was totally focused with thoughts of One and he accomplished his mission.

Nanak's Last Days

Nanak established two institutions, one of congregation of spiritualized disciples to whom he gave a special platform for performing God's service, and the other was a community of equals who could sit together, share their food, their thoughts, and similarly, the poor and rich, royalty and beggar could eat on the same platform called the community kitchen. He also stopped the terrible phenomena of *Sati*, where a widow must burn herself with her husband's pyre, when the husband dies. This led to recognition of women to have equal rights, equal status as men, in a caste-less society. He vigorously worked for the attainment of truth. He extensively sang on the virtues of truth, how to attain truth, and the path to salvation and awakening.

His main mission was to spell out the omni-oneness of God, the unity of God with emphasis on the oneness of humanity with respect and concern for humanity. Nanak stressed that virtue was gained from action and not from piety and theology. For an individual, the life's work should include three precepts. One, labor honestly (*Kirit Karna*). Second, remember God's name with divine love (*Naam japna*). This is the central core of Nanak's work. Third one was selfless service for humanity and sharing with everyone (*Vand Chhakna*).

Nanak knew his time had come and had to appoint a successor to complete his mission. The first choice would be one of his sons, Sri Chand or Lakhmi, but he picked one of his disciples who had performed selfless service. Nanak anointed one of his followers by the name of Bhai Lehna and told him, "Bhai Lehna, you have shown over these long years that you are flesh of my flesh and blood of my blood. You are my Angad, part of my *ang*, my body." Lehna was ordained as the successor and named Guru Angad Dev. Thoughtful as Nanak was, he sent Guru Angad back to his village which was *Khadur*, so that there would be no animosity or conflicts with his sons. On his last day Nanak recalled Guru Angad and informed him that it was time for Nanak to leave. He gave Angad all his writings and he walked with his family and disciples to the bank of river Ravi and sat under an old acacia tree.This village was named Kartarpur which is in Pakistan and has become the most holy pilgrimage site for Sikhs. The word was spread in the village for his last farewell. All the disciples came for the last *darshan* or meeting with their prophet. As the evening came, the followers lighted earthen oil lamps and placed them around their master. He said he was tired and requested them to sing *kirtan*. As Nanak drifted into sleep, the Muslim follower said, "He's our *Pir*, we should bury him with honor." The Hindu followers were very upset. They said Nanak was born a Hindu and they wanted to perform funeral rights with cremation. Nanak opened his eyes softly, told them both were right and he belonged to both. He told them to bring lots of flowers. The Muslims must put their flowers along his left side, and the Hindus put their flowers along his right side. And they were told to leave the flowers through the night. Whichever followers' flowers looked freshest would do whatever they wished with his body. The Guru during the early hours of the morning asked his followers to pray *japji* and drawing the sheet over himself, drifted into eternal sleep. The next morning, the flowers were checked and all the flowers looked fresh as they had been brought. When the sheet was lifted, there was nothing but the flowers. The Hindus took their flowers and Muslims theirs. The end was on September 7, 1539 AD. Guru Nanak's last message to his family and his disciples was, "Do not weep," and he uttered the following hymn(14,15);

"Hail to the creator, the eternal sovereign who hath put each one in the world to His task.

When the lifespan is run out
and the measure is full,
the soul departs the body. As the word arrives and the soul leaves
all family and friends weep.

The body and soul become separated when the days are at an end, O mother.

By thy deeds as thou acquired in the past
so hast thou received thy portion now.

> Hail to the creator, the eternal sovereign who hath
> put each one in the world to His task!
> Remember the Lord, brothers. This is the
> will, this is the way all must go."

Two Very Important aspects of Einstein that molded his life:
IMAGINATION and CURIOSITY !
CREATIVITY with THOUGHT EXPERIMENTS !
Einstein also cherished God's Thoughts and the art of Thinking

CHAPTER 2

Portrait of Albert Einstein

Introduction

ALBERT EINSTEIN WAS one of the greatest physicists known to humanity. He revolutionized physics and ushered in the era of Quantum age. Einstein was a supramental visionary who unlocked the code of gravity and acceleration. Einstein was so very much liked all over the world, not because of his genius, but by the beauty of his mind and soul that was sparkling with imagination and curiosity, and by the suffusion of his humility and finer human qualities. Einstein expressed his intuition and imagination with supramental thought processes and thought experiments. Yet, his beginnings were humble in nature. He was born on March 14, 1879 in Ulm, Germany to his parents Hermann Einstein, an engineer and a mathematician, and Pauline Koch, both of Jewish heritage from the village of Swabia, Germany. Einstein had developmental problems. When he was two years old he was not able to speak until his sister was born, who was called Maria or Maja(16). He had significant problems with language, so he preferred to think in visual imagery. He also had mild echolalia. It was speculated that he had Asperger's syndrome or autism. His sister Maria was very respectful of her brother and a very great companion. His parents then later moved to Munich, Germany where he attended a special enlightenment school called the Luitpold Gymnasium, which is now called Albert Einstein Gymnasium. That gymnasium emphasized in the study of Latin, science and mathematics. At age 9, Einstein had enthusiasm for Judaism, and composed his own hymns glorifying God. He also received religious instructions from the school. His father was extremely good with mathematics. So was his uncle, Jacob Einstein, who showed him the fundamentals of algebra and sometimes challenged him with Pythagorean theorem, and he was able to prove the theorem. Einstein thought of this theorem in pictures and triangles. He made a new proof of the theorem(16). There were two other people who stimulated Einstein's curiosity in natural science. One of them was Max Talmud, a medical student who shared his Sabbath meal every Thursday with him. Max Talmud brought him Aaron Bernstein's 21 volume series on "People's Books on Natural Science". Bernstein's writings and thoughts made a strong impression on young Einstein's mind about the speed of light and imaginative trips traveling at the speed of light on an electrical signal. The works of Bernstein opened new vistas for Einstein's imaginative fire and visual constructs(49). Talmud also helped in advancing Einstein in the field of mathematics by discussing and gifting him with a textbook of geometry. Talmud later on encouraged him to read philosophy on Kant and David Hume(16).

At age 12, Einstein excelled in mathematics and science. At this time, he also rebelled against the rituals of the orthodox Jewish traditions. He developed a profound dislike for religious doctrines and military authority with a regimented lifestyle(17).

He was probably forced to leave the Luitpold Gymnasium high school. His parents had moved to Italy and he wanted to leave Germany to avoid being drafted to the German army. Einstein's uncle, Jacob Einstein, and his father were in business together, electrifying the streets of Munich and wiring the suburbs of Munich. Jacob was an advanced electrical engineer with many patents for electric meters, automatic circuit breakers, and so the Einsteins made a major investment for a critical project of lighting central Munich. They lost the contract and their business went bust, and this had a marked influence on young Einstein, and at age 15 had a nervous breakdown and went into depression. They had to leave Germany, and went to Pavia, Italy. It provided him with opportunity to understand his uncle's work, working with mechanics of magnets, coils for making electricity, and another electrical language. So Einstein became a high school dropout at age 16. And at this age he wrote his first theoretical physics work, "On The Investigation of the State of the Ether in the Magnetic Field", where he entertained the idea that an electric current may cause the motion of the ether. He wanted to go to Zurich Polytech and he took an entrance exam, but did not make the passing grade. He then spent a year prepping at a special school where they encouraged students to visualize images. Einstein loved this Swiss educational system which allowed him to visualize his thought experiments. Einstein had great mentors who allowed him to blossom with his thought experiments.(22,23)

He was able to get in the Zurich Polytech after a few attempts. One of his physics professors, Jean Pernat, gave him the lowest grade for experimental physics and even questioned why he was pursuing physics. One of the math professors, Hermann Minkowski, called him a "lazy dog" because he never worked with his mathematics at all. But he did graduate in that mathematics class. Einstein renounced his German citizenship and became a Swiss citizen and he regarded the Swiss people to be more humane than Germans. He applied for a job and could not even get a job as an assistant teacher. One of his classmates, by the name of Marcel Grossmann, was able to locate a job for Einstein as a clerk at the Swiss patent office for evaluating patent applications. While he was looking for a job, he was passionately in love with Mileva Maric, who was also a graduate student at Polytech. Later on they had a child by the name of Lieserl, out of wedlock, and he was still looking for a job at the time. Einstein did not inform his parents about the birth of Lieserl because there was non-acceptance of Maric by the Einstein parents and family(17). Finally, he did get a job in Bern at the patent office. In Bern, there was a large Bern clock tower. He saw this clock tower everyday on his work to the patent office. He would look at the clock everyday and he was conducting his thought experiments with theoretical concepts. In his office, he dealt with questions of electrical signals and the nature of electrical integration. He used his free time to do thought-experiments with electrical signals and mechanical-electrical integration of time. In his private time, he discussed philosophy and special relativity theory. He remained in this patent office for seven years.

In 1903, Einstein finally married Maric. In May 1904, they had a son named Hans Albert Einstein. Over the years, Einstein had numerous affairs in his private life. In 1909, Einstein received a professorship from University of Zurich. In 1910, he had a second son, Eduard. Eduard did not do well. He had medical issues. At age 20, he suffered a major schizophrenic breakdown. He was institutionalized until his death. Einstein's life was not normal and caused much distress. His marriage to Maric came to an unpleasant end in 1916. In fact, his private life became more complicated. In 1919, he married his cousin Elsa, a divorcee with two daughters of her own. There is also a sad note about unknown portion of his first daughter, Liserl. Liserl, at 19 months, developed scarlet fever and was given up for adoption(16). It seems Liserl's evidence of whereabouts were not available or destroyed. Then Einstein was offered an excellent job at the University of Berlin and professorship there. This time Marie could not move to Berlin, so the marriage had to come to an end. Later on, he moved to the United States of America and he joined University of Princeton at the Institute for Advanced Study. By 1935, he decided to be permanent resident in United States of America and became an American citizen. In 1938, he was involved with initiating the Manhattan project for atomic research. The project was directed by Robert Oppenheimer and two German scientists who had discovered nuclear fission. He spent the rest of his life at Princeton, giving lectures, conducting seminars and more research until he retired from public life at age 66.

The Year of Enlightenment

Einstein spent years working on his thought platform. He was getting small moments of *Satori* or Aha-moments. But the floodlights opened in 1905, the year of enlightenment for Einstein. It seems he was in the field and drinking the nectar of existence, was riding on the photons of light. The laws of the universe were unfolding in the light quanta, and he was indeed traveling at the speed of light with zero gravity until he hit the sun and visualized the curvature of gravitational waves. Such were his thought patterns. They manifested in the language of physics and reformed the classical Newtonian physics to the modern Quantum physics.

In 1905, he completed his long-awaited thesis, "A New Determination of Molecular Dimensions," and was awarded a PhD from the University of Zurich. He was 26 years old when he got enlightened by the molecules of light. To continue his intoxication with the Unified field, he kept on pouring scientific wisdom on light, energy and gravity. He visualized his thought patterns in the language of physics scientifically. His four papers were published called, the Annus Mirabilis papers. These were published during his enlightenment in the field of physics(16,23,34). And they are:

1. The photoelectric effect, which is the emission of electrons when light meets a material. Electrons emitted are called photoelectrons. This study gave rise to quantum physics. This was actually an improvement on Philipp Lenard and Max Planck's work on the photoelectric effect and Planck's constant. Einstein expanded the significance of Planck's work. But however, Einstein doubted the wave-particle duality of light. Einstein wrote, "All these 50 years of pondering have not brought me any closer to answering the question, what are light quanta?" It is markable that he was awarded the Nobel Prize in physics in 1922 for his work on the photoelectric effect.
2. The Brownian motion – this is the motion of particles randomly in a fluid medium, resultant from the faster moving molecules in the medium. This actually gave the evidence for the atomic theory.
3. The special theory of relativity – this is the theory regarding the relation between space and time and this was based on the postulate that the speed of light is the same in a vacuum for all observations.

4. Equivalence of matter and energy, developing the famous theory of E=mc², which became the famous God's equation, and it has been used for extraordinary work.

The Crowning of Einstein

Einstein had become well-known. He had received Nobel Prize in physics in 1922 for his work on the law of the photoelectric effect. At this time he became a supramental visionary and an internationally recognized physicist. The Nobel Prize was the crowning event in his life because he had promised the prize money to his divorced wife, Mileva Maric, in his divorce settlement. Later on, in 1922, he met Neil Bohr, who received the Nobel Prize also at the same time, who was a leader in quantum mechanics(44). He met Bohr in Copenhagen, Sweden. Einstein came to see Bohr in Copenhagen, who took him via a street car. It is said in the street car they discussed quantum mechanics, science, philosophy, and the whole gamut of human thought. This all happened in the street car because they would miss the exit, go back and round-about, until they reached home. This was called the "quantum intellectual entanglement". Bohr was unable to convince Einstein about quantum mechanics. On one of these encounters, Einstein said, "God does not play dice with the Universe." Bohr would respond, "Einstein, stop telling God what to do!" Einstein objected with quantum mechanics because it did not define reality in a deterministic certainty but rather in terms of uncertainty, probability and indeterminacy(50).

Einstein was totally engrossed in the unification theory, the relation between electricity and gravitation, and a single unified theory that unites electromagnetism, gravitational field, and quantum mechanics. He worked on the unified field from 1923 to 1931. Einstein's unification would take at least a 100 years to prove it. Einstein finally accepted quantum mechanics as part of ultimate truth. The Unified field theory of one was confirmed in 2016, after the measure of gravitational waves; a 100 year wait.

Applications of Mass-Energy Equivalence Formula

In 1932, neutron was discovered, and the measurement of the neutron mass was calculated by the accelerator in United States, which produced the transmutation reaction, verifying the Einstein formula to the accuracy of ± 0.05%. This formula was used in the understanding of nuclear fission reaction, and it is possible that intense energy can be released by nuclear fission chain reaction, using both nuclear power and nuclear weapons. When energy is generated, it can be calculated by Einstein's formula. In the Trinity Test(27) before the bombing of Nagasaki, they had a yield of 21 kilo tonnes of TNT with 6.15 kilo grams of Plutonium. Einstein was very displeased and not happy about the building of the bomb and use of the bombs in Hiroshima and Nagasaki. He started a peace movement and had all his colleagues sign a declaration for peace. Einstein's formula and all his equations have been very useful in our modern day applications for internet, global positioning system, transistors, in the cellphones, and many other applications that we use everyday. His applications are also very valid for the nuclear energy production.

Later Life and End Days

Einstein was informed about his enlarging abdominal aneurysm, and that this could burst anytime and cause his death, but he decided to treat it medically and used morphine for his pain.

In April 1955, Einstein had significant abdominal pain. He collapsed at home and was taken to the hospital. While in the hospital he was writing his speech commemorating the state of Israel. This was the 7th anniversary. So on his deathbed he was writing, "I speak to you today not as an American citizen and not as a Jew, but as a human being." He never finished it. He died on April 18, 1955. It is to be noted that he had an autopsy of his brain and nothing extraordinary was found on his brain. Einstein was cremated in Trenton, New Jersey, USA, on the afternoon of his death. His ashes were scattered in the Delaware river and he did not want any monuments.

Conclusion

Einstein opened up new horizons with his creativity and power of imagination. He believed harmony was the foundation of the laws of universe. In 1914, World War I erupted and he did not support the war and was a pacifist. Einstein went against his principle when he signed the letter that went to President Roosevelt of USA requesting him funding for the Manhattan project for atomic research. On August 6, 1945, the city of Hiroshima, and three days later, the city of Nagasaki in Japan were bombed with the plutonium bomb. Unfortunately, the media called him, "the father of the bomb" because it was his initiative which started the US bomb research and it was his equation which made the atomic bomb theoretically possible." This really saddened Einstein as he never wanted these things to happen at all.

Einstein should be most remembered for his thought processes, the existence of photons and atoms, and the birth of God's equation(19,29), $E=mc^2$. He also informed us that space and time are deeply connected and it is a space-time continuum. Space-time around the sun is curving, and he had to prove it. He was a non-conformist, a revolutionary, and challenged the whole field of physics(18).

He opened the fourth dimension for us to explore. He showed that the mind has unlimited resources, and an unbelievable capacity for manipulation, contemplation and reflection. This renowned icon unlocked the secrets of atom by hacking into the unified field and exploring with his thought powers the fourth dimension to give the physics of the universe and the beauty of reality.

PART II

Quest for Truth

"Truth is what is true, or the quality of state of being true.
In accordance with experience,
Facts that conforms to reality,
Correctness and accuracy."
- Webster's Dictionary

"Cognito, ergo Sum" or "I think, therefore I am"
- Rene Descartes

CHAPTER 3

Portal of Truth

THE CODE OF creation requires whole truth and wholeness and a coherent system of beliefs. We get slivers of truth or partial degrees of truth from an inquisitive enquiry in the disciplines of physics, mathematics, metaphysics, or spirituality. It is necessary to integrate a coherent set of beliefs with corresponding facts or other pertinent correlations to have the whole truth.

To get the complete truth, it is necessary to compliment physics with consciousness, and it requires a delicate mergence of spirituality with the quantum field as it exists in nature and existence, which is a concept Professor David Bohm advanced in his book, "Wholeness and the Implicate Order." By infusing mysticism and inner technology with the unified field of one, we can experience the inner exaltation and the immersive experience of the unified filed. The beauty of this experience will become the truth to be etched on the screen of the mind. Does it require a mathematical equivalence for verification of your experience? Yes, it will authenticate it. But we do not have a formula or a model to replicate and verify an experience of beauty or consciousness. "It is what it is!"

Let me suggest that there is a harmonizing inner technology connected with the unified filed of everything contained in the vacuum space of zero point energy (dark energy). In the mystical universe, there is one reality, the un-manifest, complete within itself. The inner laws of the universe govern the outer world and the inner substance preserves the harmony and contours of our silhouette. The inner technology and vacuum space energy will be discussed in a later chapter.

Truth be told is the central discussion of this chapter and requires discussion on truth, the theories of truth, the portal of truth, and the nature of truth. This indeed would require a contemporary book on this subject. However, I will point out what is pertinent for this chapter.

There are multiple theories of truth and there are many, many partial truths that are difficult to verify because there are variables that cannot be measured and we do not have the yardstick to measure them. We can certainly measure finite things but when we move in the sphere of infinity upon infinities, both physics, mathematics and the outer complexity of our computers cannot measure them with certainty. The coherence theory of truth implies that, "To be justified is to be part of a coherent system of beliefs". This theory with the idealism of H. H. Joachim; "Truth in its essential nature is that systemic coherence which, is the character of a significant whole." Joachim also commented, "That what is true is the whole complete truth(40)."

The theories in physics are constantly being updated or replaced because they are not the whole truths. For example, mathematicians are chasing space with mathematical models and superstructures. These are mathematical grand illusions of delusions of multi-verses and parallel universes. The hypothetical postulates and models of inquiry, experiments or observations are usually not based on the totality of truth. We tend to modify or change our theories from time to time as we gather new information, knowledge or verifiable facts. For example, the Newtonian physics is not complete and is not applicable at the microscopic or subatomic levels in the quantum domain. Einstein has overruled and made it invalid for application in the quantum domain. Even Einstein's general theory of relativity took 100 years for fruition and it seems it needs new considerations, perhaps a new format and a new equation for completeness.

The portal of truth does require whole truth and wholeness to be inclusive. Physics, on its own, or with mathematics, cannot answer the void or infinity. This does require ontological reflection and consideration for completeness to satisfactorily answer the query of reality, existence and becoming. Ontology is the philosophical discussion on "study of being".

Can we satisfactorily answer what is existence, being or a thing? Aristotle may use ontology in a different way and may address being as beingness or what it is (its "whatness") and how it is (its "howness") or qualitativeness, how much it is, quantitativeness, and where it is, its relatedness to other beings". There are many ontological categories and dichotomies like essence and existence, monism and dualism, etc. that interface ontology with different disciplines of physics and mysticism(50)

Greek philosopher, Parmenides, proposed different views on existence. The first view, "Nothing comes from nothing, and therefore existence is eternal. In turn, this posits that existence is what may be conceived in a coherent system of beliefs". Indeed, if it is a truth, it is reproducible and verifiable.

In the world of mysticism and the portal of truth, there are truth-bearers. Truth becomes meaningful and believable depending on the truthfulness measure of the truth-bearers. The truth-bearers have to be insightful, believable, consistent with reality and existence. The utterances, gospels or beliefs have to be viable generation after generation; it is a process of unfolding truth in its totality. At this juncture, I would like to introduce the two supramental visionaries of this book, Guru Nanak and Albert Einstein, who are authenticated truth-bearers, where their beliefs, propositions or utterances and their viability has expressed their meaningfulness to the world at large. The correspondence theory of truth, with inclusive metaphysics, "expresses the very natural idea that truth is a content-to-world or word-to-world relation. What we say or think is true or false in virtue of the way the world turns out to be." Donald Davidson in 1986 claimed that most of our beliefs are true based on their content and incisive interpretation. Any belief or fact or content which does not withstand a process of radical interpretation loses viability and becomes an untruth."It is what it is!" This is it.

This is it!

Webster's Dictionary defines, "Truth is that which is true, or the quality or state of being true in accordance with experience; facts or reality, correctness and accuracy." Rene Descartes, "Cogito, ergo Sum" or "I think, therefore I am", the "self" is something that we can know exists with epistemological certainty. Rene Descartes was credited to be the father of modern philosophy. He started by the method of doubt and started the search for truth. So at the very start he used this methodology after ruling out all doubt and he found that beyond all doubt is, "I exist". He established it is impossible to doubt "I exist" and therefore it is a certainty. He went on to demonstrate God's existence and that God cannot be a deceiver. This was his second meditation, from his seminal work Meditations on First Philosophy. He contradicted Aristotle's philosophy with his modern system of philosophy. In his book, Principles of Philosophy, he gives a holistic view of philosophy, "The roots are metaphysics, the trunk is physics, and the branches emerging from the trunk are all other sciences, which may be reduced to three principle ones, namely medicine, mechanics and morals." Rene Descartes was convinced that universal deception seems inconsistent with God's supreme goodness. In the nature of the mind, in his Sixth Meditation, he claims, "Therefore, what I am is an immaterial thing with the faculties of intellect and will." So mind is an immaterial thinking portion of the human body. In this mind, innate ideas are placed by God at creation. These ideas can be questioned and re-evaluated or set aside at will but their internal content cannot be manipulated." He also has an understanding about God;

"Something cannot come from nothing."

God has the ability to cause the existence of the extended universe. The idea of God, something that is actually an infinite substance, namely God, must be the cause of this idea (an infinite substance), finite substance to infinite substance. God is the ultimate cause. God, the idea of a supremely perfect being, existence is contained in the essence of an infinite substance, and therefore God must exist by his very nature. This is sort of an intuitive truth. He maintains that humans are the cause of their errors and cannot blame God." For this,

Rene Descartes was blamed for what is called a "Cartesian circle". Because Descartes' reasoning seems to go in a circle, in that he needs God's existence for the absolute certainty of the earlier truths and yet he needs the absolute certainty to demonstrate God's existence with absolute certainty. According to Rene, Truth is guaranteed by God's non-deceiving nature!" I will not delve into complex propositions and mathematical equivalents to establish the truth for they cannot measure finite or infinite substances like the mind, consciousness, or for that matter, God.

God, the greatest possible being exists in the core of the ultimate existence and in reality. How do we quantify or measure the unknowable? According to Rene Descartes, "A supremely perfect being is a coherent concept." It is what it is! It is a source of truthfulness of a realization and creation. This is not a personal God we are commenting on. Physics and mathematics, or for that measure, even metaphysics does not have an appropriate yardstick for modeling the immeasurable, infinite substance. In a later chapter the Code of Creation will be presented without the complicated mathematics or physics. The portal of truth does require whole truth and wholeness to be inclusive. Life has a built-in code in the DNA of the living. The new science of epigenetics informs us that we can modify the code of genetics with our environmental signatures. In the biology of life, there is a requirement of many disciplines for completeness, reflection, and critical contemplative thinking. We rely on truth-bearers, their meaningfulness and metaphysical views. The two supramental visionaries are coherent in their beliefs and make assertions that are truthful and aim at truth. We will explore them in the next chapters.

CHAPTER 4

God

What is God?

THIS IS AN extremely complex subject based on perceptions and belief systems and yet it is an existential, perennial subject of utmost importance in the life of the universe. This subject requires utmost reverence and thoughtfulness. Yet, millions upon millions of books, articles, thoughts have been expressed and articulated throughout the ages. God remains unknown!

On a very simplistic level of understanding, one can think of God as the totality of existence, the all and end-all, the alpha and omega. From an atheist point of view, God does not exist. Then there are believers and non-believers. There are agnostics and fundamentalists. There are as many mind constructs of God as there are Gods and Goddesses to the tune of 331 million! These are purely human thought constructs.

God is One!

Indeed, all the religious thoughts and scriptures agree that God is omnipotent, omnipresent and omniscient. The credo of the oneness of God is not understood by all the conventional religions. God being everywhere in the entire cosmos and in every living creature and universes upon universes. This is God in-totality! Indeed, we construct God in the image we want. This model of God is very limiting. The Gods that religions have constructed are clothed in a doctrine and an unshakeable belief system. The doctrines, the symbolism, the belief systems and the "-isms" are all human constructs. Some of these are based on past superstitions and misinformation and not based on any verifiable evidence. Actually, most of the rituals are paganistic in origin. On the other hand, there is a unanimous calling of all religions on the concept of the oneness of God. Let us say God is omnipresent, that God is everywhere, and present in every living thing. So God exists in the neutrinos, the subatomic particles, the God particles and indeed the unified field of everything. One God for the entire cosmos, milky way and galaxies undiscovered. Can you fathom the numbers? The latest count on galaxies is 2 trillion! This goes beyond any human construct. The incredible vastness and incomprehensibility of God. This is indeed beyond any absolute measurement, infinity upon infinity. All the present systems are limiting of what is unlimited, the oneness into multiple Gods of their own choosing and earthly bound concepts. These indeed are the all concepts, systems and dogmas that are totally parochial and limited in their scope. These limitations and imageries of God served well in the past when nobody dare questioned the so-called truths. We have an evolution of God concept.

We are indeed on the crossroads of evolution in this dynamic age of information, verification and expanding social media networks. It should be noted that division and dogmas create animosities, uncertainties, and irreconcilable differences that lead to crusades and religious warfare. There is no advancement, reformation or subtle change in the core of the religious systems. The Vatican and the Pope cannot be the transformative agents. It is my impression that religions and religious doctrines with limited systems will become extinct like dinosaurs and obsolete in the near future. We clearly need a new model of understanding the modus operandi of God. In our new model, we need a greater understanding conceptually that God is omnipresent. For that understanding we must take the entire cosmic arena, universes upon universes including infinity beyond visualization. This magnanimous, grand limitless God, utterly beyond our comprehension, cannot be limited and pigeon-holed into a religious system. We need to go beyond and explain and reclaim a scientific and an all-inclusive model of understanding. We have to transcend to know the Transcendent

One, the Ageless One, the Infinite savior and God in all. We should open up a scientific and spiritual discourse based on our current level of thinking, research and possibilities. We want to be on the superhighway of knowledge and newer understanding of God particles, concept about vacuum space, zero-point energy, hyperspace continuum, parallel worlds and inclusive of metaphysics and spirituality. Theology has not even scratched into the fundamental nature of the universe and the multidimensional space concept in the quantum realm. The road to God started a long, long time ago and the journey continues without a foreseeable end. However, on the journey there are many mystical happenings, inputs of many many religions and thousands upon thousands of sects. In brief, that is the sum total of theology in the history of God. Humans are very complex and endowed with a discriminating intellect and a monumental ego system that requires them to be the center pieces with the birthing of offspring of religious divides. Then they can assert their commandments on the populace, which is eagerly groping for something to hold on to for their survival and road to heaven.

It is imperative at this stage in our intellectual development to map the progress made by both religion and science. We can review the past histories of religions and discoveries of science, both started with humans. On the religious side, there was a heavy proliferation of sects with more -isms and dogmas, without real God substance. On the other hand, science has propelled us to the vicinity and closer understanding of the workings of God. Indeed, science has excelled to the point of understanding the nature of the universe and the God particle that may govern, orchestrate, and play the cosmic drama that creates universes yet dissolves others and keeps on expanding. It seems neutrinos that have zero mass and travel at a speed of light can cause the explosion of stars. What an elegant observation!

In the new quantum domain there is further understanding, however theoretical, of the Higgs field and its messenger particle, the Higgs boson, named the God's particle, which may provide the ultimate answer of how the universe works or was created. It may not answer the eternal question of who God is. Theology needs to speed up on the super-highway of spirituality to discover new possibilities and thought experiments leading to great revelations of the mastermind, the master architect, and the lord master of the cosmos. Spiritual evolvement is advancing creativity and humanity, but religious fundamentalism is holding us to the primitiveness or paganistic ritualism and unrelenting repentance. We need to entertain God with an open mind to gain greater insights and knowledge of God creation.

We all should learn from the thought experiments of great scientists. Albert Einstein, a super physicist who pioneered the famous equation $E=MC^2$ ushered the nuclear age, conducted thought experiments that manifested his work. These were just thought experiments, and Einstein wanted to know God's thoughts. The other, Guru Nanak, was constantly exploring the thoughts of God and trying to find the supreme truth, again with his thought experiments.

We need to delve deep into the quantum realm and find the trail of God. Perhaps there is a holistic frequency domain in our universe that needs to be tapped. Spirituality should be open to intellectual constructs to see the divine and know God in closeness. In the universe, all things are infinitely connected and also interconnected on a different plane of consciousness representing the wholeness connected by infinite source of wisdom and knowledge. Please entertain the idea of wholeness and oneness. Deep down, either in the universal or religious context, we are one. We are all interconnected with God, who is also one. These concepts of oneness, wholeness and interconnectedness have been discussed at great length by scientists, saints, and seekers. According to Carl Jung, the very famous Swiss psychiatrist, "One can tap into a collective consciousness and have lucid dreams and religious visions." And according to David Bohm, a physicist, "Deep down the consciousness of mankind is one."

The sheer size and grandeur of God cannot be explained merely in our current concept. It may be a part of the larger universe, the larger picture and the working of God. Entertaining a concept with the scientific community, we can brainstorm this vast ocean of existence and break all barriers and divisions to be found in the field of cosmic interconnectedness. The world of spirit cannot be known simply by analysis or reasoning. One needs more than knowledge. Spiritual transformation and intuitive reasoning can allow the famous thought experiments of Guru Nanak and Einstein to go beyond the body of reasoning and enter the grand unified field to know the workings of God. There needs to be a new change, a new paradigm shift in our belief system with our concept of God, and our biblical thinking for creativity and advancement. If the universe is a grand unified field, then we are also part and parcel of this super cosmic field. We are all in it. This indeed is the matrix of creativity!

Not all the religious rites and rituals are mandates of God. They are all essentially human mandates to control and govern the masses. The hierarchy of gods and deities will collapse in this unified filed concept. This contains much higher realms that need to be scaled and discovered like the mountain peaks of Himalayan ranges and the depths of the Pacific ocean and the core of the earth. Spiritual transformation and the ultimate reality can only be witnessed when we cleanse the lens and remove all the distortions and misrepresentations and shed all the half-truths, fears and lies of thousands and thousands of years ago. A new spiritual world is awaiting you to scale and discover the domain of consciousness. Look beyond your beyond core beliefs and do not be booked, do not be locked in your past fixed conformity. Look beyond. Let your life flow in the present. Create new pathways for eternal unfoldment. Look beyond the beyond! Let fear vanish from your interior milieu. Remove the part of the sinner, your soul is pure! Learn the absolute truth. Remove the half-truths that were handed down from the centuries past. Understand new awareness. Be connected to the unified field of oneness and realize the grandeur of God in you!

This is the age of reasoning and relevance. There is no need to carry the burden of previous historical half-truths and lies. You have been empowered with intellectual discursiveness and reasoning. Use your power and delete the items that do not fit your spiritual and religious portfolio. You regularly make changes in your lifestyle. This is no different. Your discursive delete knob can be used to enlighten your life from the unnecessary, irrelevant, historical, unverifiable stories and events.

There is a need for upgrading ecological human engineering in our inner environment and precious spiritual landscape. We are on the edge of a precipice of an interior revolt of previous religious belief systems. It is time to take mental notes, remove irrelevance and redesign your interior with thoughtful relevance. This is your divine moment to open up and become totally mindful. The past is no more, and the present is all we have. Live in the present, and discard the irrelevant past. It is that time, that enraptured time to be in the divine and cleanse your interior from the intricate cobwebs lining your entire thinking capacity. You need inner strength and conviction to truly embark on the one god mission. You may call it your relevant calling. It is time to be one! In oneness you should think and in oneness you should bathe, and in oneness you should imbibe, and therein lies simplicity, truth and completeness of your life.

There are no protocols – no agendas or doctrines or isms or cults. You will be truly free from all the trepidations, fear and religious bondage. Just let go! Let oneness prevail.

This most powerful and most illusive simplicity will cost you nothing and you will gain and become one with the immortality of oneness!

"Father and I are one"
- Jesus Christ

How so? This statement is so very true. And yes, Jesus Christ was crucified for saying it loud and clear. The populace in his time was not ready for this concept. Even today the clergy interprets it in their flavor of thinking, allowing only Jesus to be the only son of God, and God indeed will come back for redemption and reincarnation as the messiah. "Father and I are one." It is so very true, and it belongs to the whole of humanity.

We are all part of God. Jesus Christ was indeed ahead of his times and his concept fits in with our current understanding of oneness(9).

You are graciously invited to learn and enter the realm of the new world, the world with a new global mind-shift and spiritual awakening. Open your heart, remove the cobwebs from your mind imprisoning you to the stone age. Please re-evaluate your current social milieu and your religious belief systems. It is healthy and liberating to re-evaluate your cultural and spiritual belief systems. A new age is dawning upon this beauteous civilization. A newness is surging in all the regions of the world. Listen to the call of your inner dweller, and pay heed to the stirring in your consciousness. A rising consciousness will empower all humans in its path.

A new age of spiritual transformation is upon us so we can transcend our programmed life and dissolve our social conditionings and karmas. We have been conditioned to be sinners and contained within the four walls of our patterned, programmed mindset. It is time to break the shackles of our bondage and claim victory. It is time to say, "My soul is pure. It never sinned." It is time to review your boxed up life. It is time to do away with the lesser understanding of yesteryears and elevate consciousness to the waves of light and grace that will quicken spiritual awakening and a new global mind-shift. Be prepared to bathe in the river of light that will usher us into the golden age. Prepare yourself for the new galactic orientation with ensuing cosmic events and cosmic alignment. Open your heart and mind, listen to the call of your entrapped spirit, for a new happening is taking place. Be open and receptive to receive the new coming. The new understanding has a potential of sweeping the entire human psyche, the global world view, the religious systems, and the belief systems so that we render asunder the whale of ignorance and live a life without sin, without persecution, and without condemnation. We live a life embedded in liberty, happiness, love and greater understanding of one god. So be it.

Einstein's Religion and God

Einstein was born as a Jew and had significant Jewish heritage. During his early childhood, Einstein had a special interest in Judaism and observed kosher dietary habits and Jewish religious holidays and Sabbath. He even composed hymns about the glory of God. At age 12, Einstein was fascinated by mathematics and science and started rebelling against religion and stopped observing orthodox practice of the Jewish religion(17)

Later on, Einstein was branded to be an atheist and he did not like the labelling and he instead would get angry and tended to be critical of atheists. Einstein was asked the question if he believed in God. Einstein did not believe in personal god. He questioned such a god who judges our actions. He considered God who is of a much superior power who reveals the universe and is an infinite spirit (16,21,24)

Einstein had a deep religious instinct. All his scientific fervor was actually cultivated by cosmic religious feeling and the source of all his thoughts. He had his own views on science and religion. He believed science can only be advanced by scientists who are seeking truth and greater understanding. He also believed science and religion are mutually beneficial .i.e, one cannot understand without the other.

Einstein was totally against a deity and a personal god and religious rituals and class distinction, and he did not believe in the Jewish concept of God. It has been asked of Einstein if he prayed. He probably did not pray to a personal God, but believed in a superior Spirit that exists in nature.

Einstein was deeply engaged in his thought experiments. He was constantly, unceasingly invoking the mathematics of God and endlessly carrying on in his thought experiments, the intersession of the unified field imploring to give him the solution, the equation. Actually he was constantly praying in the unified field like a scientist would!

Guru Nanak's View on Religion and God

Guru Nanak made it very clear that he was not a God or a saint. He was basically a teacher, preaching the message of God. He was born to

Hindu parents of a higher caste. However, all his admirers, followers, disciples called him various names like the Avatar, Nanak Lama, Hazrat Nanak Shah, Satguru Nanak Dev, prophet of the people, and one of the spiritual writers, Mehrban, wrote, "He does not look like a man of the world – to our good luck, we are meeting in him God himself."

Nanak declared early on that he was not a Hindu or a Muslim. He was against rituals, the caste system, and during his time ritualism was rampant. During one of his travels, a group of yogis invited him to join the religious sect. These yogis were exercising their rituals and their practice. Guru Nanak responded,

> "Religion lies not in empty words.
> He who regards all men as equal is religious.
> Religion lies not in wandering outside
> to tombs and places of cremation,
> nor in postures of contemplation,
> religion lies not in roaming abroad,
> nor in bathing at places of pilgrimage.
> To live uncontaminated
> amid worldly temptations
> is to find the secret of religion."

On the spiritual path he made sure he did not identify with any one religion. He was constantly invoking the power of one god and was engrossed in the thought patterns of Ek Om Kar. He spent a lot of time totally absorbed, reflecting on the cosmic patterns of God. He would sing and recite with humility and compassion and at the same time align with cosmic vibrational consciousness, embracing One in a rapturous communion. He had mystical experiences. He formulated that the ultimate One is the path of unity consciousness and he lived in the ocean of Oneness until Nanak became the embodiment of oneness.

On one of his famous travels to the Himalayan mountains near the sacred Mount Kailash, a spiritual dialogue took place with very advanced yogis. This was recorded as the Siddh Ghost, which is included in the

Sri Guru Granth Sahib. The Siddh Ghost is a dialogue which consists of 73 verses(13). These yogis believed in specific systems of philosophy with their external symbols and rituals. There were many questions asked of Nanak. We will try to answer three pertinent questions Nanak was asked.

1. What is the origin of life?
2. What faith predominates the time?
3. Who is your guru, whose disciple are you?

Nanak replied:

"The breath is the beginning, from the air originated
life. My system also belongs to the time of origin.
The true guru is the Word and the Soul is its
disciple (this is the Surat-Shabad yog).
What keeps me in my detachment
is meditating on the unutterable gospel. To the one divine word
God is made real to us
and the disciple controls the mind."

Guru Nanak's preamble to Japji Sahib, which is the Moolmantra, encapsulates Guru Nanak's views about God. The complete Japji consists of 38 stanzas and it has become the divine morning prayer of the Sikhs. Moolmantar:

There is One God,
Ek Om Kar, the supreme truth,
the creator,
beyond fear,
beyond rancor,
timeless omnipresence,
beyond birth and death,
self-existent,

>   obtained by divine grace.

It finishes:
>   God, the eternal, was truth in the beginning
>   is truth through all ages,
>   and shall be truth forever.

Guru Nanak's life was a prayer, a composition of God's essence, a reflection of the thought of One, Ek Om Kar (2,3)

He was a genuine teller of truth, clothed in the vesture of truth and singing the poetry of truth. No wonder he was in the vineyard of truth. Such was Nanak's religion and his unshakeable conviction in:

>   Ek Om Kar
>   Sat Naam!

This thought is embraced by the 31 million souls who practice it and is blazing with the unification of oneness of God and humanity.

PART III

From Physics to Consciousness

"To taste God
create a God-thought
with the thinking molecules
whirling in the womb of existence
bathing in the nectar of
God consciousness
                                        in Ek!

CHAPTER 5

The Unified Field; Thought Experiments and Divine Singularity

THERE HAVE BEEN astronomical advances in particle physics, neuroscience, epigenetics, metaphysics and mysticism. We tend to separate them to different disciplines and branches of knowledge. However, on the individual and planetary level there is a holistic synthesis and harmony. This is indeed a global sea change in our understanding of consciousness and quantum domain.

From a lifestyle perspective, there is a growing global awareness of unity and diversity. We are embracing the quantum domain and the unified field. It is not surprising that quantum physics is proving scientifically that everything is interconnected with the unified filed of existence and the oneness of existence. Quantum physics confirms consciousness creates reality. Consciousness is fundamental and a metaphysical mystery. This may blow your mind that consciousness is not created or contained in the human body and it is not a local phenomena. This is a new paradigm shift that Nanak and Einstein elaborated on in their thought experiments and quest of one. The inner quantum soul is all light and photons that carry the message of consciousness. This is basically the new inroad in the physics of spirituality. This actually should be a hallelujah moment!

You have thinking molecules. The light is within. All the thinkers, prophets, and now the scientists are saying look within for deeper inner experiences.

The universe is an integral whole. This concept of wholeness and oneness was the incessant, repetitive, and fundamental message of Nanak, and in it is the Holy Grail, "Ek" or One, is embedded!

Even Jesus Christ said, "My father and I are one, and the kingdom of heaven is within." From a scientific perspective, David Bohm, professor of theoretical physics at Oxford University, England has written at length in his book, "Wholeness and the Implicate Order". He says, "Science is demanding a new, non-fragmentary worldview, in the sense that the present approach of analysis of the world into independent existent parts does not work very well in modern physics. It is shown that both in relativity theory and quantum theory, notions implying the undivided wholeness of the universe would provide a much more orderly way of considering the general nature of reality(41)."

The totality of existence, the totality of life, the totality of humanity is a prescription for living coherently, cohesively and interconnectedly for the greater well being and respecting the integrity of our planet.

The theories of Galileo, Newton, and Descartes viewed the universe as inert without consciousness and obeying the laws of motion. The new physics, the new leaps in neuroscience and cosmology have a very different worldview. Newton's laws of motion are still valid at the macroscopic level but they break down at the microscopic level. Einstein could not work with Newton's laws. He had to modify them. Einstein's new physics ushered in a new era of quantum mechanics. All the disciplines may focus on different aspects of modern physics and particle physics at the quantum level. Interestingly, the new biology and neuroscience have also changed our phenomenal worldview. There is a common thread that weaves a cosmic web. All interactions in our universe at subatomic levels create new interplay with some sort of intersession of sorts that brings in separability and convergence in a way that the moving parts reciprocate certain degrees of mutual interrelationship. This produces the consequential coherence with harmony and unification. This then creates unification on all separate disciplines, thoughts, particles, into a coherent correlated wholeness. And finally, an inherent wholeness. That is the new worldview – an intelligent, conscious, self-aware unified field of oneness and everything. In this grand unified field, everything is connected to everyone with energy or vacuum space—zero point energy, as they say in quantum physics. In other words, we are all floating in consciousness, connected in the vacuum energy, and at the core of the universe everything is connected. This indeed is the new physics of spirituality and the neuroscience of consciousness in the great disciplines of cosmology and existence.

There is a point of "unknowns". That is when physicists and mystics take us to the theatre of theorems, postulates, mathematics and hypothesis. There is a logical methodology to pursue our hypothesis. The unknown is the enigma of the soul, mind and consciousness of God that has not been deciphered and decoded like the DNA of life. They are not physical entities. In essence, they all belong to the unseen, image-less, genderless unified field of oneness. We do not know how thoughts are processed by the mind or who indeed is a thought maker? A thought essentially is a quanta of information and energy and flows in the endless stream of consciousness. We cannot explain all these complex processes by the billions of neurons firing in the brain. Yes, we have silos of massive memory banks' storage facility in our brains which we can recall to give the subjective of any complex human emotions that we encounter. This we call, an emotional experience. Indeed, life is a tapestry of human experiences of joy, sorrow, love, beauty and expectations; we all live for experiences.

We do not as yet have a scientific model to measure the unknown entities of soul, consciousness or mind. We can measure energy, mass, gravity, electromagnetism and speed of light mathematically using Einstein's models. If we cannot see soul or consciousness or mind, it does not mean that they do not exist. The quantum physicist can find infinite intelligence in the universe. Computers, no matter how sophisticated, to date have no soul, mind or consciousness. They cannot produce the biologic proteins like neuropeptides and neurotransmitters that are produced by the brain to experience the emotions of love or the longings of the soul, mind and nature. Maybe in the new quantum computers with highly complex algorithms, we could crack the code and install the field of consciousness and intelligence.

For humans, the faculty of mind has reached a very high level of sophistication and can advance to infinite levels of Nikola Tesla, Einstein, Buddha or Nanak and other millions of humans engaged in the study of existence and nature of reality. I wonder what was the modus operandi of these masters and visionaries and how they directed their lives for the greater good of humanity.

This book is an attempt to penetrate the thoughts, minds, life and philosophy and methodology of two supramental visionaries, namely Guru Nanak and Albert Einstein. They were very simple human beings but they fashioned their lives, their purpose and their eventual existence in a certain way. Guru Nanak became the founder of a new vibrant, energetic religion (not by his choice) with 31 million followers worldwide. And Albert Einstein became the founder of modern physics with worldwide acceptance and application.

So what was special about these two supramental visionaries? It was certainly not the capacity of the brain size. Einstein's genius brain was smaller by 15% than an average human brain size. It was definitely not their physicality, for they were not the most muscular giants. There were no outstanding physical traits or features that can account for their massive accomplishments. You can say the same things about Jesus Christ, Nikola Tesla, Picasso, Lord Buddha or Prophet Muhammed. All of these, one could conjuncture were "god sent", but they were human beings like you and me. If there was such a phenomena, we are all children of one god and god does not discriminate. And they had the same DNA code of life. Since we do not have a special 'god-sent' model of inquiry, we will examine the human model of inquiry and understanding. Actually, that is the most practical methodology and it is applicable for all to use, for these two visionaries gave us the way, the code of creation, and the tools of their monumental achievements. They deserve all the accolades and praise and recognition for their revolutionary methods of achievement of their goals. They did not have any special physical devices or apparatus for examining their goals very successfully. They had the same faculty of the mind with imagination, intuition, intelligence and intention that you have, however with different degrees of development or use. Their imagination, their vision boards, their intuitiveness, and perhaps their intellectual manipulation of the mind was of a much higher order of thinking and connectivity with cosmic consciousness than ours. Let us try to delve into what is known about the unified field, mind, consciousness and theorems and formulations. The worldview of a person depends on one's preconceived beliefs, the development of the perceptive apparatus and the faculty of the mind, the immaterial [inaudible] faculties of intellect and will. According to Rene Descartes, "Innate ideas are placed by God at creation. These ideas can be questioned and evaluated at will but their internal content cannot be manipulated." We have the advanced thinking apparatus with innate ideas waiting for a mastermind to flower them in the realm of all possibilities. After the inception of ideas, we cultivate them in an incubator of our brain. These ideas blossom with the tincture of time and grow through the process of idealization, imagination, and design, with excellence or perfection through the visionary lens. When a polished idea or thought is conceived, it needs the determination, the intention, persistence and an unbending perseverance for the thought pattern to become the brainchild of the master goldsmith. Such is the deliberation and celebration of legends in the making. They go the extra mile with their deep thoughts, with the uncompromising resilience of a mastermind. Nanak and Einstein were made of such particles of mastery and created their patented indelible signatures, flowing in existence for generations to inherit their treasury of thoughts and code of creation.

What is the Unified Field?

All the laws of nature display quantum connectedness with oneness, at the core. Unified field was there before the Big Bang. Einstein displayed that there was grand unification of all the fields of electromagnetism, gravitational fields and energy fields. It is also inferred that unified field, "reverberating within itself", is the generator of all the laws of nature and all the critical forces exercising control and managing the whole universe. With exquisite simplicity, all the laws of nature are under the command, or as Nanak would say 'Hukam', of the supreme singularity and embodiment of the unified field. Indeed, there is perfection in the execution of the order of the universe. The order of the universe or the Hukam in every particle, in bees and butterflies, in humans and biodiversity, in neutrinos and electrons, in galaxies upon galaxies. This order or Hukam is the will of Ek Om Kar. The supreme self permeates and pulsates in every atom of existence with the order embedded within. The code of creation is inclusive of the matrix of creativity and energy of the universe, the unified field with infinite intelligence and consciousness.

It is as if Nanak became one with the source of Om, pulsating awareness of Ek Om Kar, and Einstein formulated the energy of creation, the vital universal life force. Professor John Hagelin, a physicist from Stanford University, California, USA has summarized unified field in a way that is very understandable for all without elaborate mathematical equations and particle physics. He has validated the link between macroscopic and microscopic, science and spirituality, and is building quantum bridges between consciousness and unified field. Hagelin has also elaborated on the quantitative correspondence between pure consciousness and the unified field. He does claim that pure consciousness is the unified field. He also set up a classification of physics into five different levels; level 1 being the unified field, the core; level 2, Supersymmetry; and level 5 including electronics, telecommunication, computer science, mechanical science, aerospace technology, and laser technology. Level 3 on his table is chemical, nuclear technology, quantum mechanics, and atomic physics. All these are connected to the fundamental core of the unified field. Our new understanding and description of the unified field according to professor Hagelin:

"Unified field is an ocean of pure existence,
an ocean of pure abstract intelligence, at the basis of universe....
This ocean of intelligence is steaming
with unmanifest energy, it is reverberating within itself, so dynamic,
infinitely energetic, that it is
erupting in itself. It is called
vacuum energy or zero point energy."

Can you believe this is the power of Einstein's code combining the unified field in an elegant simple equation, $E=mc^2$.

Divine Singularity

The scientists are informing us that the unified field is the epicenter of the field of consciousness and intelligence. Everything radiates from this field of existence. Pure consciousness and intelligence have structural and functional correspondence to the unified field in certain vibrating frequencies of photons, electrons, or graviton. Pure consciousness is more subtle than thought patterns. Indeed, the vacuum space is also very subtle.

Einstein introduced us to the fourth dimension of time. He then introduced the speed of light as the cosmological constant. Later on, he also had to introduce the concept of space-time continuum. In physics, space-time fuses the three dimensions of space and the one dimension of time into fourth dimension. Einstein's famous god equation $E=mc^2$ used the c^2, the speed of light, as the cosmological constant.

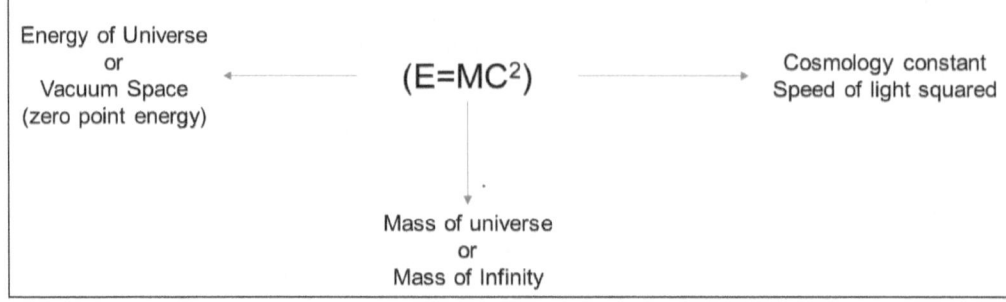

Figure 1. The God Equation

The self or *atma* is the fourth, the undivided inner dweller. When we pay attention and rest in the *atma* (soul) we activate the light within. The light that is within, *Atma* or self is the seer within. Guru Nanak calls this the fourth state and this is the *atomic* particle which is the divine cosmic constant. When you achieve a state of Self awareness, you are anchored in the stream of light, consciousness and quantum connectedness.

"At this deepest level of meditation (*Naam Simran*), *atma* is activated and you are connected. *Atma* is subtler than subtle essence and exists in the fourth state, and the fourth state is realising oneness with the self, the eternal! This is like living in a light matrix, riding on the electromagnetic wave. This light matrix vibrating in the electromagnetic frequency connects with Ek Om Kar. By communion with the word (*Naam-shabad*), you can attain the status of a Siddha, Prophet or a mystical knower.

Truth (*sat naam*) is the one and true is the Holy name."

Quantum physics confirms consciousness creates reality. Consciousness is all around us and we are tapping in it. One has the ability to connect at a higher thinking level and information level. This is what supramental visionaries do when they connect the finite to the infinite. When you are in the light matrix, you transfer information or thoughts at the speed of light energized by the vacuum. This is the way of Omni-oneness.

Indeed, the unified field of Einstein to Steven Hawkins can be understood by what Professor John Hagelin informed us, that unified field is ultimately the field of consciousness. When you connect, you can become infinite. This is called 'vacuum engineering' in modern lingo, as was experienced by all the mystics and spiritual beings. You can experience enlightenment as you enter in divine singularity of "Ek Om Kar Sat Naam". Let us also remind ourselves what thoughts can do.

> "We are what we think.
> All that we are
> rises with our thoughts.
> With our thoughts,
> we make the world."

- Lord Buddha

CHAPTER 6

The Code of Creation

The Code

This is not about the act
of bringing the universe into existence.
Not about, "In the beginning was the Word, and the Word was with God, and the Word was God."
It is more ancient than the
Ancient one. It is beyond the beyond!
It is about Oneness
and how to be One!

THE EXISTENCE OF humanity is a gift from the creator. This is a precious gift that we should be grateful and share it for the greatest good of our planetary inhabitants and planetary wellness. Some of the messages are encoded in scripts, encoded in the DNA of existence, and in the DNA of life that allows us to entertain all possibilities. The code of creation is a prescription of sorts that reveals the matrix of creativity.

Nanak and Einstein have unfolded new codes in the spiritual and quantum domain. They contain the essence and beauty of this complex web of cosmic dimensions. Combining Einstein's coded unified field and Nanak's code of the core of one, we have the code of creation. We can harness it to get the master key of creation, the vortex of power, the grand treasury of the vacuum that never empties! The code of creation combines these two powerful creative forces of nature to give us the key to unlock the treasury of the unified field with the codified signature of Guru Nanak as you perceive Ek Om Kar pulsating in the cosmos.

Figure 2. The Signature Logo of Guru Nanak

Historical Perspective on Codes

Everything is dependent on coding the code, the QR code, the computer source codes and the ability of computer systems to read the codes and express the message in clear, understandable ordinary language. So you need decoders and specialist programmers to create very sophisticated algorithmic codes.

What is a code?

A code is a set of letters, communicable message, symbols, principles or laws that are codified information systems. Computer codes are coded in secret cryptic language and numbers. There are specific coding systems. Often times coding systems are used for transmitting secret messages where the ordinary language or arrangement of data is converted into a coded message. We used to have the Morse code for long distance communications. Hidden in these coded transmissions are technological and creative powers. There are codes from time immemorial, starting from the ancients, the Brahma code, the Bible code, the Moses code, the code of life in the DNA and genetic code. It seems all scriptures have codified messages but they are not mathematical formulations.

The codes have been there from ancient times. The Upanishads are the mystical and philosophical texts of Hinduism and Vedas, and in them are coded messages. The incredible "Om" comes from the Upanishads. Om is the icon that has universal connotation. It was first mentioned in the Upanishads as a cosmic sound. It is indeed a mystical cosmic sound. Om has evolved over time for initially being a meditative mantra to the sound of the universe, the symbol Om as is written in Hindi (see fig.1) is encoded as the symbol of consciousness and most sacred mantra of *Brahman*, the ultimate reality in Hinduism. It also refers to soul (*atman*).

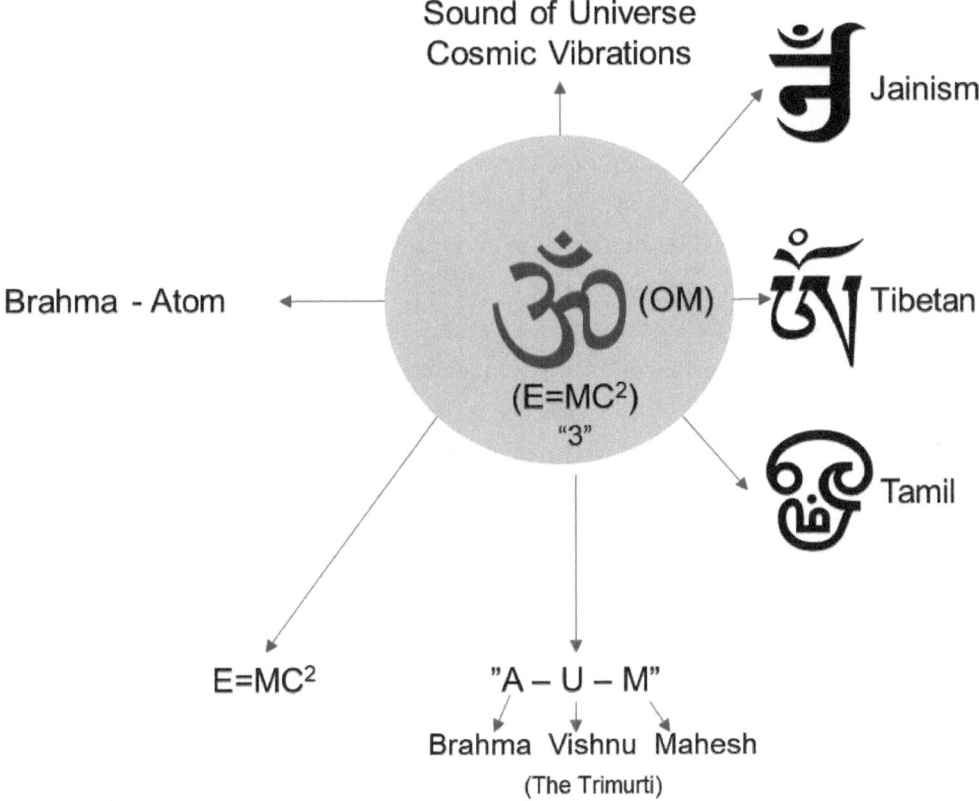

Figure 3. Significance of Om

It is now collectively, regarded as the primordial sound associated with the creation of universe from nothing. Indeed, Om has evolved to the highest stature in Hinduism:

> Om is *Brahma*! (ultimate reality)
> Om is all this! (observed world)
> - Upanishads

Om is also linked to *atman* or soul. In the Maitri Upanishads, the text claims that Om represents *Brahma-Atman*. The sound is AUM, which is the body of soul. Lord Buddha also recognized the power of Om which is used frequently in Buddhism. The famous Buddhist complex Om is called *"Om Mani Padme Hum"*. This six-syllabled mantra or encoded format is repeated in every Buddhist temple everyday and has six empowering aspects that help perfect the essence of Buddhist teaching; for perfecting the practices of generosity, pure ethics, tolerance, perseverance, concentration, and wisdom. All this is encoded in *Om Mani Padme Hum*! So the Buddhist code is the totality of sound, existence, compassion and consciousness.

So when you recite Om, you awaken your awareness. When Om is resounded, it fires the thought of awareness within to know the thought of God within and realize the soul or the higher self. Even Bhagvat Gita uses Om as the symbol for the indescribable, impersonal *Brahma*. The Puranas (medieval Hindu texts) represent the Om as the Hindu tri*murti*, Brahma, Vishnu and Shiva. So Guru Nanak represented Om in a different light without the tri*murti* effect to the absolute monotheistic unity of God and called it Ek Om Kar! Om Kar is the Om-Maker. The symbol Om has universal applications and is indeed linked to different aspects of God like:

<u>Om</u>nipotent
<u>Om</u>nipresence
<u>Om</u>niscience
<u>Om</u>nioneness

Om is different in different linguistic patterns and forms, depending on the religion or country where it is practiced (see fig.1). There are indeed thousands of names of God who is essentially one. Om has become a sacred spiritual utterance of many religions like Hinduism, Jainism, Buddhism, and Sikhism.

Guru Nanak was immersed in the name of God for God has no name, no gender yet is always masculine, has no image and is indoctrinated in religious practices. Few and rare ones are aware of Nanak's code which is enshrined on every stanza of the Holy Granth Sahib, the holy scripture of the Sikhs. Yet the code has been there and has not been decoded or cracked. If Nanak's code is amplified and applied, it can become the master key that unlocks the greatest mysteries of the universe and reveal the code of creation. Similarly, the Bible code was only recently cracked by three Israeli scientists and mathematicians that unlocked divine messages hidden in the Bible linking Biblical messages from the dead sea scrolls in the Hebrew alphabet. For this book, the words Omkar and Oankar are interchangeable and have the same meaning with different connotations. However, Ek Om Kar is very different from Om or Omkar. Moreover, Guru Nanak's Ek Om Kar is a coded encrypted signature logo, that calls for deep meditation and decoding. Guru Nanak's iconic symbol which is on the cover of the book is also very unique, absolutely different from the tri*murti* of A-Brahma-U-Vishnu-M-Shiva. It is Ek or one and not three. There is no duality or plurality in this singularity. Guru Nanak's symbolic logo has a subtle difference which characterizes infinity enfolded onto Ek Om Kar. There has been much debate about the pronunciation, spelling, dialect, whether it be in Sikh dialect or Vedic text. The symbol Om is different in different dialects, regions or countries as misinterpretations frequently happen(fig3). If one analyses the singular thought pattern of Guru Nanak, there should be no misunderstanding. The unique logo formulated by Guru Nanak is Ek + open '*Oora*', the very first alphabet of *Gurumakhi*. The open *Oora* has an open and an extended end pointing towards infinity. This is the encoded *Oara*. It is infinity unfolded. Just one (One) Oh! Ahh! Whatever way you wish to pronounce or write or interpret, it is your change but it won't make any difference. However one should look with the lens of Oneness of Guru Nanak to get the total meaning of oneness. So in this book, Omkar and Ek Om Kar or Ek Onkar or Ek Oan Kar will be regarded as same with different dialectical writings in text and pronunciation.

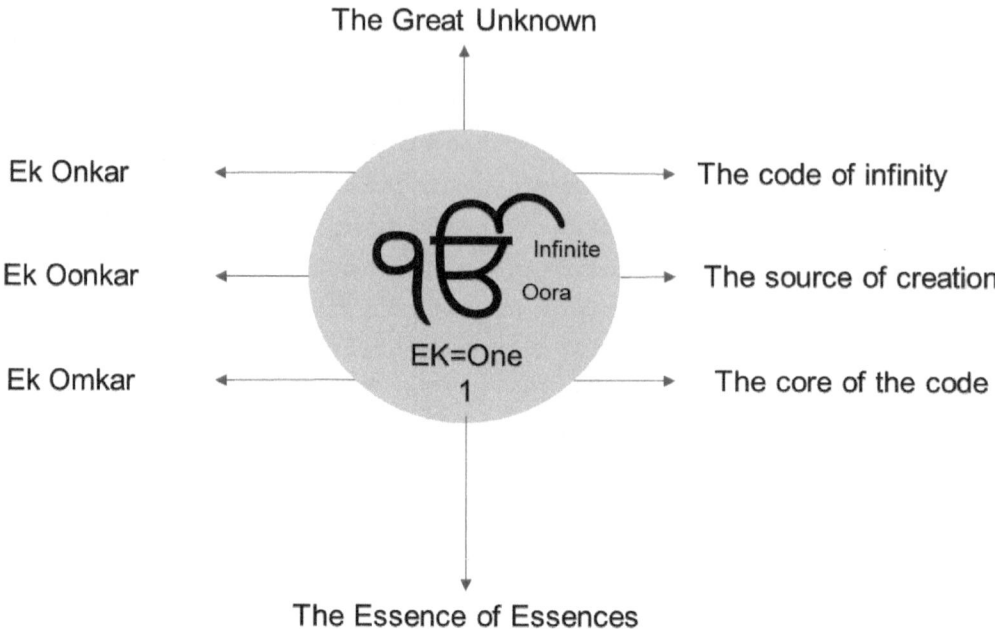

Figure 4. The Gateway of Transcendence

Guru Nanak's code is contained in his scripture logo which has been written in different dialects and means that there is only one reality, only one God, and it was called Ek Om Kar or Ek Oan Kar or Ek Onkar, and also Aumkara. This has caused some misunderstanding for pronunciation and meaning. The Sikh scholars and philosophers have vetted and embraced their dialectical prowess on their mental screens (see references). According to Indologists, there are only dialectological and syllogisms that are not consistent with Guru Nanak's dialogue. To further the understanding of Guru Nanak's message, we will entertain his dialogue which is labelled *Omkar Bani* or *Ramakali Dakhani* which is in Sri Granth Sahib (pages 929-938). We have our limitations and understandings which only Guru Nanak can correct and Guru Nanak's truth prevails. Guru Nanak was on a journey to south India and arrived at a temple in central India, named Omkar on an island. The island resembles the Hindu symbol of Om and is called Mandhata on the Narmada river. This island is divided into Shivapuri in the north and the southern portion is divided into Brahmapuri and Vishnupuri of the trinity. Shiva is regarded as the supreme being on the island. There is a central worship of *Shivalinga* called Parameshvara. Guru Nanak was astonished by the polytheistic and paganistic worship of the *shiva lingum* at the Omkar temple. He had a discourse with a pundit of the temple. Guru Nanak corrected the Vedic interpretation and cleared the confusion to the pundit about God. He said there was only one God, not a trinity or tri*murti*. There is no room for duality on this plane of existence. He inserted Ek into Omkar, that God is an eternal entity without form. A section of the *Omkar Bani* translates thus:

> "From Omkar Brahma was created.
> From Omkar was created mind and spirit.
> From Omkar time and space were created". (page 929-938 Adigranth)

Omkar is the world of liberation. Reflect seriously on the word Om. The word Om is the essence of three worlds. Open it, gather the gems and pearls of Omkar, then thou shall get the real diamond of Omkar.

Om is the cosmic divine sound, the vibratory medium of God to create galaxies, the thought-forms, consciousness and the Unified field. Actually, Om is the medium of quantum physics, and the cosmos is a pulsating wave. Om is a pulsation in the cosmos. In many scientific sessions and discussions, we have been informed the whole universe is a vibrating pulsation. Ek Om Kar is the maker of Om. Ek Om Kar used the sound, OM the vibratory frequency of Om to cause the singularity event of Big Bang.

Om has also been likened to the famous God equation of Einstein, $E=mc^2$ (see fig.2).The symbol Om as written in Hindi is: $E=mc2$ as a mirror image By decoding Ek Om Kar, you will observe God is a master singularity, the cause of all singularities, for the creation of superstrings upon superstrings, the supersymmetry, the 2 trillion galaxies, the embodiment of omni-oneness in the unified field of everything, including the vibrating pulsing medium of Om.

It behoves us to be in Ek Om Kar and infuse in the ocean of oneness and become the pulsating wave of Ek Om Kar! There is no religion here. There are no Gods or personal deities here. Even Nanak or Einstein are not present in this code. It is all one! Go for this transformational awakening and superconsciousness.

Just look within
and
dissolve in Ek Om Kar to become One

To become supramental metahumans go on the platform of one with the code of creation and awaken your awareness. The code of creation (CoC) (see fig.5) is the protocol of truth, consciousness, and totality of oneness in combination with the god equation, $E=mc^2$ times infinity squared.

$$\text{COC} = \text{EK OM KAR} + \text{E} \times \infty^2$$
$$\text{EK} = \text{One}$$
$$\text{OM} = mc^2$$
$$\text{E} = \text{Energy of vacuum space}$$
$$\text{Kar} = \text{Creator}$$
$$\infty^2 = \text{Infinity upon Infinities}$$

Figure 5. The Code of Creation (COC)

Where E is the energy of universe, m is the mass of the universe and c^2 is the cosmology constant, the speed of light squared. E is the vacuum space of zero point energy. If you want to know beyond the beyond, it is infinity squared.

The final equation will be forthcoming from the field of existence, perhaps by a new set of supramental metahumans. By formulating the code of creation which is inclusive of creativity, communication, and consciousness and infinite intelligence in the Unified field, one can harness and tap into the field. This can happen if one is open to unbounded awareness and awakens to the super Unified field of consciousness and Oneness.

We are told that a split second after the Big Bang singularity, a galaxy was born with the milky way and thousands of stars with planets orbiting them in our galaxy. Our sun is just one of about 200 billion stars in our galaxy. That is the immensity and vastness of one galaxy. When there are 2 trillion galaxies, we have infinity upon infinities. Mathematics and all of physics and cosmology cannot fathom this final equation. So help us, Ek Om Kar.

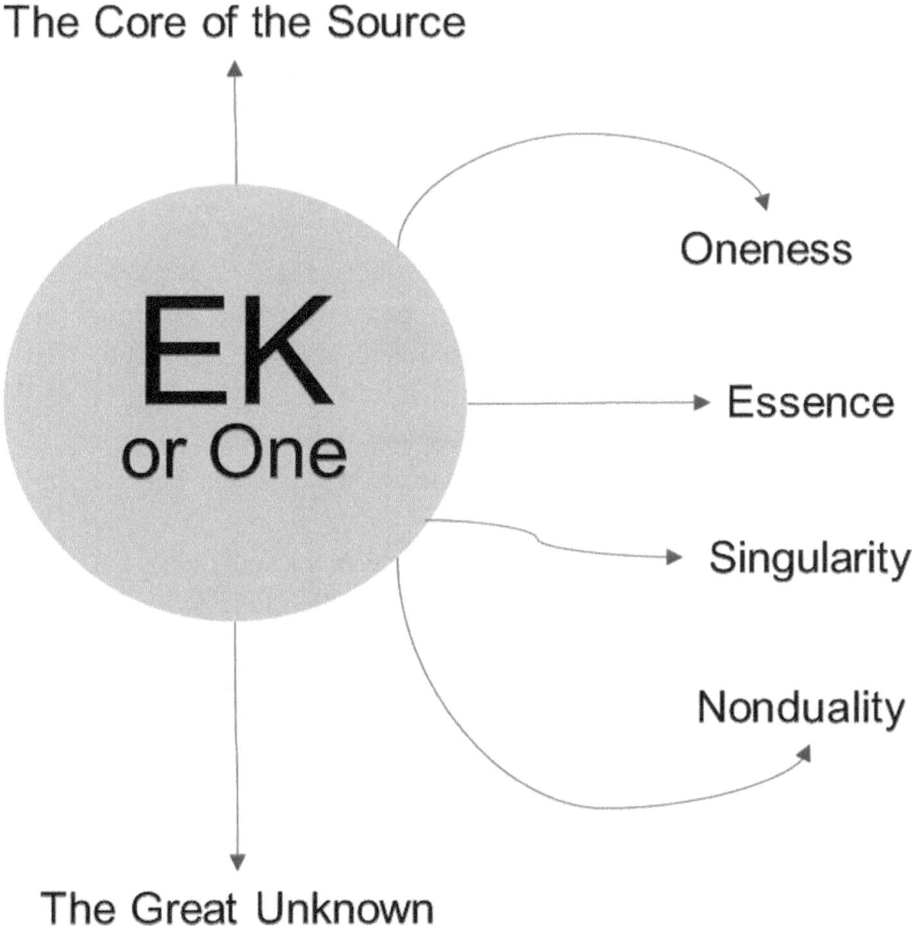

The Core of the Source

EK or One

Oneness

Essence

Singularity

Nonduality

The Great Unknown

Figure 6. Ek or One is here, there, everywhere. We are in the matrix of EK.

For us, the code of creation is simple and has immense practical applications in our daily lives, in the planet's survival and recognition of the quantum field and divine singularity. These practical applications and developments of supramental states will be presented in the next chapter.

This universe is Ek Om Kar's laboratory which helps us to experiment and discover our inner self and explore existence and experience reality. This is the place where you awaken and experience your inner self through self realization. This is the realm of unknown, unmanifest adventures and discoveries.

This is a thought experimentation laboratory where Nanak and Einstein prayed ceaselessly on their thought experiments and awakened the divine force to carry out their message and mission and the profundity of their supramental visions. They entertained congregations of thought patterns of light, energy and Ek Om Kar.

Guru Nanak spent more than 30 years of his life contemplating and reflecting on the thought patterns, on the true name with a focussed singular attention, never deviating, never doubting. Such was his conviction, and he gave us a very simple message: Ek Om Kar – Sat Naam!

Albert Einstein worked on thought experiments by riding on the speed of light for more than 16 years and he gave us a very elegant, simple equation: $E=mc^2$.

Such is the intention, attention, perseverance, unbending fixity on thought patterns and experiments in the unified field of Ek Om Kar and vacuum space of these supramental visionaries. We can develop a supramental state by entertaining the superconscious phenomena of simplicity, 'I am One!' and unfold the new signature of the field of consciousness. This is a way to know the unknown, and to understand the secrets of existence. The code of creation is the connectivity link for your dimensionless mind to get a taste of infinity and awesomeness of the universe around you. When the mind meets with the infinite timeless Eternal Being, infinite possibilities bubble up from the field to discover the extraordinary moments of epiphany and most cherished experiences.

You are on the cusp of a breathtaking awareness that will design your future. You can ride on the light of One and become: I AM ONE!

>   I am one!
> I am one with the wind
>   with the breath
> I am one with the atom
>   with the heart
> I am one with the universe
>   with the non-separability
> I am one with the code
>   with the philosopher's alchemy
> I am one with Einstein
>   with the $E=mc^2$
> I am one with Nanak
>   with Ek Om Kar
> I am one with existence
>   with the field of consciousness
> I am one with God consciousness
>   the essence of essences!

CHAPTER 7

SUPRAMENTAL Beings and Inner Technology

EK (One)

> Ek is the needle and the thread
> that weaves the fabric of the cosmic web
> into a wholeness of eternity.
> Ek is embedded in Ek Om Kar.
> Ek is the Holy Grail.
> And
> $E=mc^2$ is the vital life force of existence.
> Together Ek & E=mc2
> move the wheel of existence
> in the matrix of a
> seamless Oneness!

EK OM KAR is a creative symbol of power, strength, and magnanimity. It is the gift of Guru Nanak to the world. This, along with Einstein's gift of the code of the Unified field, contain the code of creation. In the code of creation is contained the mystery of existence, creative fiat, the space-time continuum, and the vacuum space with zero point energy. Everything in the universe is a part of a continuum, which means one seamless total whole, according to David Bohm, a former protege of Einstein and professor of theoretical physics at Oxford, United Kingdom.

The inner technology and the quantum field, along with the alchemy of the mind, one can incorporate power thoughts with thinking molecules to achieve the realms of transcendence and obtain radical possibilities with the code of creation. This is the Nanak effect. You can expand your inner technology by engaging in the meditation of the soul (*Naam Simran* with *Surat-Shabad yoga*). This is the inner practice to achieve the state of Nanak consciousness through "*Shabad*" consciousness. To achieve this state of creativity and awareness, one anchors in the unified field. This is the Einstein effect and field consciousness. This is absolutely mind-expanding, opening radical possibilities with the unified field. You cannot separate the two, the Nanak effect and the Einstein effect. At the core they are one. Separability is an illusion.

> Nanak was a trader of truth,
> a connoisseur of truth,
> the commodity was truth
> the gem of Ek Om Kar.
> Nanak uttered the name of truth.
> Nanak spoke the language of truth.
> Nanak sang the song of truth,
> thou, thou, thou,
> Theraa, Theraa, Theraa,
> Ek Om Kar, Sat Naam,
> the distillate of truth.
> Nanak was a pure truth carrier
> of Ek Om Kar
> and
> truth-bearer
> of the region of truth called: Sachkhand!:.

Einstein searched for truth
  in the mind's laboratory, created thought experiments,
    he rode on the speed of light,
      followed by the trail of truth,
        in the field of truth,
          the Unified Field,
            always
searching for God's thoughts,
  in the field of truth.
    Einstein was a truth-bearer
      of the Unified field.
        Nanak and Einstein drank
          the nectar of truth.
            And they were memorialized
              in the house of truth.
                Ek Om Kar!

            It is not Nanak's truth
              Nor is it Einstein's truth.
            Truth is free from human chains.
            Truth is not contained in
                    preconceived ideas.
            The truth cannot be used for exploitation, for
            enslavement,
                for denomination,
                for annihilation and personal
                aggrandizement.
                We are disciples of
                    TRUTH.
            In truth we live, not in the image of
            personification.
            In truth we experiment
                    our thoughts, ideas and ideals.

In truth we seek happiness
In truth we believe.
In the garden of truth there is no hell or heaven !

The Significance of EK (One)

Ek means one, the lowest whole number, numero uno, being a cardinal number like one dollar bill. Ek is as ancient as the beginning of speech. To explore the domain of Ek (one) and the manifestations of Ek, we need to understand the original existence of one and the codified set of principles contained within the manifold aspects of one.

According to Merriam-Webster's dictionary, one is the ultimate reality, the central source of being. For Guru Nanak, Ek is the alpha and omega, the entire wisdom and understanding of the universe. He meant the whole universe is contained in Ek. Everything, everyone, everywhere is Ek, the totality of oneness. Ek decodes the truth of the Trinity of:

>   Father, Son, Spirit,
>   Brahma, Vishnu, Shiva,
>   proton, electron, neutron,
>   substance, essence, nature,
>   heart, mind, body
>   cosmic creator, preserver, destroyer,
>   theist, atheist, agnostic.

Ek is the foundation of oneness and is the most basic understanding. Ek is the engine of connectivity. Ek connects all religions, all the inhabitants of biodiversity, the multitudes and universes upon universes. Ek has the transactional and transformative ability which is beyond one's imagination! According to Heraclitus: the One is made, of all things and all things issue from one."

Our world is fragmented and torn between doctrines and dogmas, and as a result of this, suffering has multiplied in our fast-moving world with commercialism, exploitation of the earth and endless wars. Once we recognize that Ek or One is the Holy Grail embedded in one God:

"My father and I are one!"
- Jesus Christ

The Holy Grail indeed is Ek. All the scriptures agree that there is one God! This is of utmost importance in the implementation of our monumental task. I wrote a book entitled One Word can Transform Humanity and Save Our Planet, and in it I explored the concept of Omnioneness. This embraces the singularity of one, within is the ocean of oneness, and oneness of God.

Ek is the particle, the wave and signature code of creation. When you do the meditation of Ek, and embrace Ek, you slip into the core of eternity in a rapturous communion with divine love. In this Nanak rejoices, and the nectar of Ek saturates you.

Two famous physicists have theorized that there is one electron. Robert Feyman and John Wheeler, two of the best modern quantum physicists, entertain the idea that the whole universe is a single electron because all the electrons are the same weight and that one electron traverses time and traverses and bounces in time countless times.

Karl Pribram, a very respected neurophysiologist at Stanford University, and David Bohm, an eminent quantum physicist, believe that the universe may be a giant hologram(42). A hologram is created by interference pattern. Light waves can create interference patterns and form images called holograms. For more details, refer to the Holographic Universe by Michael Talbot. And I have discussed this on the concept of a superinfinity hologram, in The Book of One, the Eternal Testament. Let us believe that the cosmos is a grand, giant hologram and we are part of the hologram. A hologram is three dimensional and the holographic film contains all the complete information of the whole. So we are all bits of the grand hologram, and this is where we can explain that God is omnipresent in the creation, in every atom and in each one of us. I will take the liberty to introduce a unifying concept with modern understanding of consciousness and quantum physics. Let us assume, that the whole universe is a superinfinity hologram, and at the quantum level the whole cosmos is one Supreme Electron. This would satisfy Robert Feyman, John Wheeler, David Bohm, and Leonard Susskind's holographic reality and a single electron theory. Indeed, this will take a lot of convincing and a lot of experiments to confirm this idea.

Actually, this might even be helpful in the concept of Ek Om Kar, that there is one creator, one supreme electron, and in it is everything, everywhere, in everyone, and that is the one superinfinity hologram and we are all bits of that superinfinity hologram; we call it God. This would also fulfill Einstein's Unified field theory of everything. Nanak's main core concept was unity in Ek. At the core, there is interconnectedness of everything. Yes, Einstein was working on a Unified field of everything and indeed it has become the Unified field of One!

Ek is a supreme particle and signature code of creation, the ultimate supreme electron with superinfinity hologram with endless creativity, infinite consciousness and unlimited possibilities. Duality and illusion have no room to play in this totality and singularity of Ek.

Supramental Beings

Supramental beings are humans like Guru Nanak, Albert Einstein, Nikola Tesla, Jesus Christ, Prophet Mohammed, Lord Buddha, Leonardo da Vinci, Shakespeare, and many many thousands and thousands of humans who have contributed and benefited the greatest good for all humanity. Supramental refers to a comparative state of mental thinking beyond the normal level. This is the higher dimension of a transcendent being in their awakened and self-aware state. This is the radical transcendent state that draws from the fountainhead of consciousness. This activity of mental thought is achieved when one is in the awakened state and in touch with the higher self which is called "*Atma*".

Supramentalization is the ultimate stage of Nanak's "*Surat-Shabad yog*" and transformation of the thought processes with ultimate union with the "*Shabad*". This is the transformation of the individual to a higher level of thinking, drawing from the field of pure consciousness. This indeed is the royal path of connecting with the source, the timeless infinite. This is the state of awakened self-realization functioning in an effortless way.

I have written on these new evolved humans who have reached the supramental state through deep concentration, persistence, intuition and imagination. This is not about divinization of the soul but achieving the awakened state of self realisation. This state of supramentalization can also be achieved by the simple techniques used by Einstein, a seeker of the truth, with thought experiments, a mind capable of curiosity, imagination and hyper-focussed with a set of thought patterns and an immense capacity for persistence and perseverance. If such is the burning desire and intention, then the universe will itself, in its totality, respond and reveal.

The portfolio of inner technology consists of:

a) heart, mind, body coherence
b) meditation of the soul with *Naam Simran*
c) selection of thoughts, creating thought patterns
d) self-realization, or the awakened or self aware state
e) the infinite you with infinite possibilities.

The Course of Events That May Unfold

1. The path of the divine. It may require three days and three nights of continuous meditation of the soul to have that awakening that Guru Nanak had.
2. The path of mathematics and scientific experiments, with thought experiments that Einstein underwent for 16 long years.
3. The path of truth to understand spiritual knowledge and laws of nature and spiritual truths.

4. A state of self awakening, superconsciousness and spiritual enlightenment. The key to realization is through meditation of the soul with the essence of God. With meditation of the soul, one realizes and unfolds the unfocussed thought of the transcendent God.
5. A manifestation of the laws of physics through thought experiments and validation through applications and practicals.

Heart-Mind-Body Coherence

To achieve a state of oneness, clarity and attentive focus, there has to be total coherence in the body temple. This forms the template and the platform to achieve the status of supramental beings.

What is Coherence?

To be coherent is to have the quality of wholeness for core goals, so that the heart-mind-body are unified for achieving life's purpose. When the two hemispheres of the brain are synchronized and become coherent, the brain's cognitive powers and capacity improves tremendously to give the Aha moments of pure genius. The bi-hemispherical synchronization can be accomplished by a set of exercises that ensure that the two hemispheres of the brain are coordinated and in sync for peak performance. These exercises are used for the initial process of activating what you desire. You can synchronize with mindful meditation at the alpha level in your brain's meditative state. Next, one has to have coherence in the brain's electrical circuitry. This is a process achieved through meditation, yoga, and also new technology involving psycho-genesis. The next step is to declutter the mind from its endless internal dialogue with random thoughts darting from all directions. Delete whatever is unnecessary, specially the belief systems that may be ossified, closing your mind and encircling it with dogmas and doctrines. Letting go is a very useful and important step in decluttering and opening the mind space for the new focus mind shift required for total body coherence, attention and awareness.

At this juncture, remember we are in 'Vacuum Space' of the Unified field that is bubbling with consciousness and infinite intelligence all around us. This is indeed a new concept, that astrophysics and cosmology have opened, this new creative vista. Some may even refer to this phenomena called a simulated reality. But I think this is a false reality hypothesis. The body is where the heart and mind reside. The body is the temple where the self or *atma* is. So remind yourself of the importance of the body to be in a totally relaxed state and in communion with the whole. Be centered. For a breathing exercise, be centered on your breath. Observe your breath going in and filling your lungs and the chest expanding with oxygen. The breath is one that serves the whole body and connects you with the infinite. The breath is a ligature to your finite body to the infinite and oneness. This breathing exercise and awareness, just breathing awareness, will focus you to your higher self. Sometimes we call that focussing on the third eye between the two eyebrows. This is where sometimes it's also called the third eye center, and where resides the pineal gland.

We are in a timeless vacuum space without any preconceived illusions, navigating the Unified field of pure stillness, pure consciousness without the thought ripples from before. So you have become empty of any preconceived thoughts, emotions, beliefs, and dogmas or doctrines. When you develop keen focus on your third eye and you are breathing and following your breath, there is no room for random thoughts to disturb you. Keep on this zone of awareness until all thoughts, until all ripples settle down and still be in a totally relaxed state from crown to toe. You are creating a timeless song with self-awareness riding in the unified field with pure consciousness and with no intermediaries. Only the timeless, dimensionless self and a thought free mind bathing in the medium of light filling up and expanding into your heart, mind, body, and suffusing it with ever expanding light. Remember, your soul or your A*tma* is a light matrix. When you are in this moment of stillness where there is no thought movement, indeed the heart-mind-body complex have become an efficient functioning whole and there is total coherence. This is about the process of self-realization to the awakened state connected to the source of the unified field. This prepares you for the awakened opening state.

Coherence in nature and coherence in biology are the foundation of our connectivity in the grand field of oneness. There is connectivity held together by the exceptional coherence clue in the diverse fields of quantum biology, consciousness, quantum physics, and the phenomena of non-locality. Let us examine a quanta of energy. In their natural, unexamined, unobserved state, quanta are not in one place at any one time. A single quanta could be here or there or everywhere in space-time continuum. The phenomena of non-local for (action-at-a-distance) is the ability of objects to instantly know about each other's state, even when separated 1 million light years. Non-locality is what Einstein called, "spooky actions at a distance". This was described in his famous EPR paradox(see glossary). This contravenes the principle of locality, the idea that distant objects cannot have direct influence and only immediate surroundings can influence it. Einstein was totally upset by the conclusions of non-locality and he truly did not accept it until his dying days(50).

Non-locality happens due to quantum entanglement. In entanglement, particles interact with each other and are dependent on each other's states and properties and lose their separateness. These particles in quantum entanglement are not separate. They are all one, united, interconnected. Bell's theorem in 1964 was a profound discovery that non-locality and quantum entanglement phenomena are real, as shown by experiments. Einstein and his colleagues, Podolsky and Nathan Rosen, in 1935 performed the famous EPR experiment (thought experiment) to testify to the quantum non-locality (22,23,44). To get the details of this revolutionary experiment, please refer to Einstein's works regarding the EPR experiment. We cannot go into the details in this short summary. According to this experiment, a separation does not exist even with billions of miles away. All the particles are entangled. This phenomena of quantum entanglement is at the centre of all quantum field discussions(22). Particles, when they are entangled with each other, are not only in one place or at a distance place, but exist simultaneously in all measured places at the same time. This quantum non-locality is not affected by separation either by a millisecond or millions of light years. In the field of particle physics, there is an observation of non-local connectivity called teleportation. The experiments that confirmed that teleportation exists between individual particles and also atoms. Teleportation experiments have given us new insights that ancients mystics used to perform *Atamic* teleportation to distant stars or even in astral teleportation. So at the quantum teleportation of entangled particles, one can create incredible speed of transfer of information by the next generation quantum computers.

Humans are also a giant coherent quantum system in perfect dynamic equilibrium and in a state of homeostasis. One billion cells performing trillion biochemical reactions and the human brain with trillions of synaptic actions comprise this giant coherent human enterprise. One single human being created from a single zygote. All systems coherent for optimal functioning.

Einstein discussed the issue of consciousness and likened human being as part of the whole, called the cosmos, a part limited in time and space. Humans experience thoughts and feelings as being separate from the universe. He called this a kind of delusion that limits us (31). Consciousness, as we currently know it, is everywhere in the vacuum space. Indeed, the entire unified field is consciousness. It is important to understand Vacuum, which in English means empty space. But in cosmology, it has very different implications. At the basic level, it is cosmic space without matter. This has been called the blue sky. We all know the space is really not empty. It is filled with all kinds of information. There are all kinds of wireless electronic signals, information packets present in the space. All our electronic gadgets and data are transmitted through this apparent empty space. The ancients believed it was ether. After experiments, the concept of ether was modified to absolute vacuum. Einstein used the three dimensional space in his general relativity theory with time to make it the four dimensional space-time continuum. The vacuum space has been continuously modified as our understanding of cosmology and astrophysics is evolving. It seems the vacuum is not only active, but it transports light, energy, information, and links particles, atoms, solar systems, entire galaxies and consciousness. So there is coherence and connectivity throughout our 2 trillion galaxies. The information is carried by charged particles with their rotating packets of electrons and positrons that are carried by vortices of energy at inconceivable speeds of 1 billion times the speed of light, i.e. 10^9 c (c = speed of light), according to the Russian physicists, Shipov and Akimov.

So the information carried by the charged particles in spin by the vortices interact with other vortices to create interference patterns, i.e. the wave impulse in the vacuum medium intersect and interfere to become wave fields and carry information or the state of our galaxy. The information carried is non-local and propagate as cosmic hologram. You will recall that a hologram is formed by the interference pattern created by two beams of light, especially laser beams giving us the 3D image of the object.

This information is present throughout the universe or the Akashic field. Indeed, the quantum vacuum is the omnipresent, all existence pure consciousness and all intelligence and creativity. It is as subtle as can be, beyond our perception like prana, the life energy force, which is the womb of existence, the force of everything and the foundational medium. We all are in it! This is the authenticated, verified knowledge from the quantum vacuum by all the current cosmologists and physicists. At the mystical level, there is unity between the quantum vacuum and A*tma* or the soul. When Nanak used to ask his companion Mardana to bring his R*abab*, the stringed instrument because he was getting "*Dhur kee Bani aayee*", the celestial song of the universe is dawning on him. He literally meant it and now we know he was in touch and connected to a higher level to the quantum vacuum, the core of the supreme divine force. We call it God, Ek Om Kar, the One and only One.

Guru Nanak was singing the divine message, the celestial song of the universe. It came to him in compositions of stanzas and divine poetry. In modern language we would say he was manifesting the thoughts of God. The supreme atom forms the quantum vacuum with the thinking molecules. The atoms within can contemplate and think and process information as the universe is unfolding.

Let us engage in the playground of the universe with a new set of tools for empowerment and experience the multidimensional worlds with infinite possibilities.

CHAPTER 8

The Royal Path

It is That:

>That which is beyond
>>the ordinary,
>>>that thinking spark of
>>>>creativity
>>>>>that knows no separability
>>>>>that which is within
>>>>>>waiting to be awakened.
>>>>>>>It is That I Am
>>>>>>>>spaceless
>>>>>>>>timeless
>>>>>>>>causeless
>>>>>>>>fearless
>>>>>>>>deathless!
>
>I AM in the light matrix, the pure soul!
>In One is the mystery of all life
>>The One for all
>>The One in all
>>the One we seek!

THE ROYAL PATH is an extraordinary path for manifestation of your defined, focused intention. This path requires no intermediaries. It is devoid of personal gods or deities. There are no religious dogmas or doctrines. It is not affiliated with any political or religious organization. It is based on the Portal of Truth and the

Physics of Consciousness. This path is a synthesis from the works of Guru Nanak and Albert Einstein. The Royal Path represents the utter newness, filled with thinking molecules sparkling with Oneness. The meditation of the essence with the true name, unlocks the doorway of existence embellishing radical possibilities with the Code of Creation. The Royal Path requires significant preparation and the steps are as follows:

1. Heart, Mind, Body coherence
2. The Alchemy of the mind
3. Mind-Spirit coherence (with *surat-shabad* yoga)
4. Meditation of the essence
5. Transcendence with the code, reprogramming and rewiring
6. Radical possibilities with the code of One
7. Mindful meditation of One, the essence of the transcendent.

Internalize your thinking with auto-suggestion, affirmation, visualization or even self-hypnosis that: I am One, I have all the resources of the Unified field with infinite possibilities empowered by the mastermind of Ek Om Kar and the code of Creation!

The Unified field is the field of all possibilities and pure consciousness. This is your playing field. In here you dream like Einstein and Nanak, because it is the field of creativity and unlimited potential. To access this field one needs to embrace the light of the inner dweller, the soul or *atma* which is the guide and knower of the field. The soul, light, *atma*, the higher self is in fact the light matrix, the pure essence that is the "*Shabad*" of Guru Nanak. It is also the word of the ancient scriptures called:

>In the beginning was the word
>the word was with God
>the word was God.
>According to Nanak,
>the word is the *Shabad*,

the true name.
The *Shabad* is the light embedded in all.

Again, according to Guru Nanak, in the *Siddh Gosht*, "*Shabad*, God's word abiding in all, is Guru; *Surat*, or mind attuned to Guru (God) is disciple." This is the *Shabad-Surat yog*.

The *Shabad* or the word is the connecting link to the field of all possibilities. This is the quantum soul electron you want to excite and activate that will access the jewel of the essence, the true name. This is an unparalleled method to harness your intuitive largess with immense abilities. It does not require the mention of any gods or any deities. Also unwind the wheel of karma and be relieved of all so-called sins. Dharma and karma imprison you and are the great resistance to your ultimate freedom. Break loose from the bondage of karma and sin, for your soul is all light and pure. Break all barriers to enter oneness or EK. This is your breakthrough. Go to your interior, to the sacred sanctuary. Dive deep and go beyond, beyond the strictures of your beliefs, all entanglements, and be free right now.

Right now, right this minute, shed your self-limiting garb, the ossified shell of religious dogmas and deities. Completely detach yourself from the past rituals and practices; remove them like you change your clothes. It is as easy as that to be free. Then you are on your way to unleash your potential in the field of all possibilities. When you are prepared and ready and your heart-mind-body are coherent you become totally mindful of one, I AM ONE, one within, one with your family, one with your friends, one with the planet, one with the word or *Shabad*. This is the preparation for the meditation of Ek or One.

Bring to your mind the quality of awareness and mental shift that is required to create your reality. Be super motivated! Design your high value high priority thought patterns by using the faculty of your incredible mind. The alchemy of the mind can fashion your thought patterns with a concentrated, focussed, dominant thought which has the capacity of a very powerful weapon that pierces through the mountain of obstacles. Next, believe in yourself. A resolute mindset with unbounded conviction and commitment infused with an indomitable will makes you the master of your destiny, the architect of your life.

Simply put: I AM ONE. Your Heart Mind Body are synchronized with your creative thought patterns. You have empowered your thought pattern. Dream, imagine and be curious like Einstein to master your vision, your choice, and be unlimited with infinite potentialities. You design your life with your thought patterns. Visualize your thought patterns. Use your mental alchemy to formulate your thought patterns. Implement your idea in your mind. This is your creative genius loaded with passion, intention, imagination, intelligence and intuition. Please disengage from all negative thoughts. Delete all your core negative files of "I can't", and again release and eliminate all your fears and negative thoughts of feelings into the mind's bin which you can delete. If you want reinforcement of your thought patterns, write your core ideas, goals and create a thought folder which you have to upload in the form of dimensional thought patterns or visual images on the screen of your mind. Stick with one most important thought pattern from your thought journal. Circulate this thought pattern in different colors on your mind's platform. You can use the rainbow colors to visualize it till it gets implemented. Keep on improvising it, editing it until it is crafted pitch perfect, to your liking, a clear vision of what you want emerges on your mind screen.(Details of this technique will be forth coming in a new book The Ek Suite.)

MINDFUL SOUL MEDITATION

Preparation for this meditation is of utter importance.

First and foremost, you need clarity of purpose, power of concentration and focus of total intention. This mindful soul meditation or meditation of the essence of Ek Om Kar is the highest practice and should be performed with reverence to your soul. Recognize that the Essence is contained in the timeless One, which gives the spark of realization and allows you to anchor into the Unified field. You want to pitch your vision, your life-plan in your creative thought patterns. This meditation will allow the whole brain functioning synchronously in a coherent way with heart mind body complex.

In this meditation please leave all your baggage, your past mindset, all previous beliefs in the mind's bin and press in your mind, let go and delete. This will let you dissolve the previous mindset and thoughts. You now have silos of mind space in your subconscious mind to upload the new program that you want to have for your life. The reprogramming requires rewiring your software with new algorithms and thought patterns in your subconscious mind. Please make sure there are no outsiders in the new you, not any father figures or gods, not even Nanak or Einstein. This is a clean slate and an ultra mind with the Ek suite.

The Ek Suite

The Ek suite is a gift from Nanak and Einstein, the two creative forces for manifestation and enlightenment and power thinking for your future. The Ek suite is the knower, creator of thoughts and all of knowledge. It contains the deepest level of knowledge from the vacuum space or the Akashic field of the unified field. In the cosmic theatre of existence, the orchestra of heavenly bodies plays with an unseen conductor; a vibrant orchestra of light particles and waves that create a symphony with unstruck cosmic music. In the theatre of existence where light particles dance and collapse into waves to the unstruck music from the orchestra of the stars creating universes and collapsing others in the black holes. The fabric of the Cosmos is weaved by the electron of EK. The vacuum which is the womb of creation is the glow of quantum entanglement for non-local connections everywhere in space and time and transmits ions linking all non-material and material particles instantaneously at bosonic speeds. The quantum vacuum is the library of the universe. It is also called by the ancient mystics the Akashic cosmic library. It is said everything is recorded in this cosmic library from the beginning of time, to the timeless future on the dimension of time. Edgar Cayce, the famous American psychic, and Nostradamus, the French physician and reputed seer with his 942 poetic quatrain, predicted future events. Both of them accessed the Akashic library with their inner technology and meditative practices. Ervin Laszlo, a well known philosopher, wrote in his book, Science and the Akashic Field, "The Akashic field is real and has its equivalent in vacuum space, zero point energy. This zero point Akashic field is the subtle sea of fluctuating energies from which all things arise: atoms and galaxies, stars and planets, living things and even consciousness. This zero point Akashic field is the constant and enduring memory of the universe. It holds the record of all that has happened on Earth and in the cosmos and relates it to all that is yet to happen." (52).

According to Swami Vivekananda, a great Hindu mystic, "Akashi is the omnipresent, all penetrating existence. Everything that has form, everything that is the result of combinations, is evolved out of this Akasha. ----- It cannot be perceived; it is so subtle that it is beyond all ordinary perception."

This new understanding for modern physics and the science of spirituality essentially unites different paths, different equations into the medium or the matrix of creativity. Zero point energy is all around us and it is a sum total of all energies including the life force, called *'prana'*. The vacuum space in the unified field is the dynamic energy filled vacuum. The vacuum is also the Sanskrit *Akasha*. This matrix has been given different names. Just like God has a thousand names. It is still one and only one, Ek Om Kar! Ek Om Kar defines all aspects of one God, without any compartmentalization, dualities and *maya* metrics. Ek Om Kar contains all of *Akasha,* all of Om, all of *prana*, all of energy, all of these and more in one, without any religious iconography.

Let us embark on this experiential journey of truth. There are different paths and techniques for achieving your goals and you can follow your bliss wherever it takes you or you can take the Royal Path, the direct path without any intermediaries.

The Ek suite (detailed in another book)

1. For optimum experience there must be coherence of heart-mind-body complex and total body relaxation from crown to toe.
2. Have access to 31 minutes of uninterrupted time, essentially away from all electronics, preferably in a dark room with a lit candle. Silence is of utter importance here.
3. Everything rides on your breath; the oxygen, the photons, the electrons, the thoughts. So be absorbed in following your breath going in and breath going out. Become aware of your breath going in, which is followed by a physiological pause, and I call it space of Ek or One. And breath going out carrying your carbon dioxide, your mind's bin emptying out. Follow the breath. It has four phases, the new breath going in, the pause, the old breath coming out, followed by a pause, and the cycle completes with breath going in. Breathe naturally with your diaphragm going down during inspiration and letting go during expiration. Let go of your redundant thoughts.
4. Bring back your focus to your breath awareness and start emptying, empty, empty, empty all your thoughts. This is a great emptying, making room for the vacuum and mind space.
5. Go into deeper meditation. As you become aware of your breath, you automatically switch into alpha state of your brainwaves rather than the normal beta brainwaves. To bring more concentration, focus on your so-called third eye centre, the pineal gland located behind your eyebrows. You can initially look at the candle flame with your loop of breath awareness or look within to the midpoint between your eyebrows with your eyes shut. This will cause some strain to your eyes but with practice, you will get used to it. Try to maintain looking with your eyes pointed inwards to the midpoint between your eyebrows to the third eye. If you get tired, let go, relax with your breath, and then come back turning your eyes to the third

eye. Next, pay attention, full attention, and bring your loop of breath, the breath going out and breath going in to your mind's eye, the third eye centre. Breathe from your third eye when you are comfortable. Introduce Ek or one to your breath in the space of Ek. When you breathe in, drop Ek in the space or pause or the gap. With full focus and attention and concentration you have opened a seat of I Am That! Ek vibrates on the threshold of this centre.

6. With the next step, you open your heart centre with love and joy and happiness. Create the ambience of gentleness and a state of eternal happiness so you can receive the grace from the seat of the soul or the light matrix which transmits the soul current, the amp of Oomph! This soul current opens the corridor between your heart centre and the third eye centre. In here, you present your most valued asset, (life-plan) the most delicate presentation of your thought pattern with total clarity and conviction. This is your asking with your life plan. The universe will respond on its own accord. Be patient with gratitude and love.

7. When you are ready, you can go on a transcendental ride on the wings of Ek to the region of truth and beyond, decoding the code of creation for the ultimate mystical union of your soul (*atma*) to one (Ek) in the fountainhead of Ek Om Kar!

EPILOGUE

The Quantum and Spiritual Domain.

WE HAVE COME to understand that consciousness is all around us. We are floating in the matrix of consciousness. The universe responds to the mind. With our ideas we create the world. The mind interacts with cosmic consciousness to create your reality. Indeed, consciousness is everywhere in the universe. We humans have a cosmic dimension.The Universal Mind has thinking molecules contained in the subatomic particles.

We have come to recognize the Unified Field theory. The universe has become the playground, bubbling with creativity, infinite intelligence and pure awareness. The super vacuum that is all around us is holding our cosmos and connecting with the medium. Galaxies upon galaxies and everything in existence is connected through this medium. This immeasurable vastness is beyond our thinking when you consider infinity upon infinity. This is the expanded view beyond our observable universe. Einstein's theory of relativity does not apply here. Just like Newtonian classical physics cannot explain what quantum mechanics can. We are indeed in need of physics and mathematics (404) to explain all the unexplained categories of unknowns present in our universe. We need a new super equation to explain the pathways to infinities. We will need new Einsteins' and Nanaks'to further unravel the mysteries of the universe.

The mystics, the spiritualists and modern day agnostics do recognize that at the core we are all connected and are in the vacuum space with perhaps zero point energy. That being said, the quantum physicists, the astrophysicists and particle physicists also carry the impression that at the core we are one, the powerful realm of One!

I do not have the philosopher's stone, or the supernatural force to unfold what is at the core, the source of everything; to understand this exceptional oneness, I call it Ominoneness which is absolute oneness. This omnioneness, this immeasurable vastness is pervasive here, there, and everywhere. If we venture to explain omnioneness and omni-consciousness we begin to time travel the imageless, timeless, dimensionless, unending, ever-expanding, eternal, immortal, beyond infinity, beyond eternity, this massive vastness, Oneness! All pervasive in the whole cosmos, in you and the smallest god particle! Such is the grandeur and supremacy manifesting in all the universes. What is it?

What is it? A wonderment, one supreme electron! Indeed, a supreme electron of such unknowability, immeasurability and immensity exuding cosmic consciousness and infinite intelligence, pure awareness and Absolute Oneness. This is the one we should explore and contemplate and reflect on. This is the platform of One. This is the omni-computer that has all the codes, the super secrets, super intuitive and probably unhackable!

One supreme electron is a giant holographic wave in the matrix of dark matter, super massive black holes containing super gravity and infinite super vacuum energy that is holding our cosmos and everything in existence filled with cosmic consciousness and intuitive intelligence. In quantum entanglements everything is connected, separation is an illusion.

The Avatar of the mystic world and the physicist of the Unified field of one were traveling with thought experiments and thought waves, curating thought patterns in the field and praying ceaselessly, with absolute fixity of purpose, never doubting but always curious and imagining, never daunted by failures or rejections. Such was their mind empowerment! They left an indelible imprint that can unlock all possibilities. Remember, you can use their code of Creation for all radical possibilities! They were the lords of manifestation, manifesting the One, manifesting the Code.

Ode to the Code of Creation

Floating in the stream of
cosmic consciousness
flirting with the matrix
to get mystical experiences,
I dive deep within,
to seek profundity
and explore the creator
of the code
embedded within
my soul's codon.
Dancing in the whirling atoms
chanting Ek Om Kar and calculating $E=mc^2$
a formulated supernatural concept lies beyond,
embedded in the code of Creation.
A fleeting spontaneous mystical state
awakens a deep sense of
Oneness;
a truth has been transmitted
soaring in the heart of truth
imbibing the molecules of truth
the pure vineyard of truth,
embedded with luminosity of light.
Ascend the vast stairways of light to the
ecstasy land for a transcendental ride
on the photons of light, and be blended
in the Supreme Light. Herein eternity unfolds.
This is the ultimate mystical union of your
inner being to your divine lover.
This is the *Surat-Shabad yog*.
This is the Essence of essences,
the grandeur of One, on the celestial throne,
streaming with love and light

and total consciousness
singing Ek Om Kar
Sat Naam
$E=mc^2.$

This is the platform of One—the Code of Creation !
Drink you the blessed one,
        the immortalizing molecule of One!

BIBLIOGRAPHY

1. SRI GURU GRANTH SAHIB: 4 VOLUMES; Translated by Gopal Singh; 1978; World Sikh University Press, Chandigarh, India.
2. Iqbal S.: The Essence of Truth, Japji and other Sikh scriptures; Allen McMillan; 1986; USA.
3. GurBachan S; Talid, Japuji: The Immortal Prayer -Chant; Munishirar Manoharlal; 1977; Delhi, India.
4. Puri, JR; Guru Nanak: His Mystic Teachings; 1982; Radha Soami Satsang Beas; Amritsar, India.
5. Khushwant Singh: Hymns of Guru Nanak (Translations); 1978; Orient Longman; New Delhi, India.
6. Kushwant Singh: A History of the Sikhs; Oxford University Press; 1987; New Delhi, India.
7. Wazir Singh; Aspects of Guru Nanak's Philosophy; 1969; Ludhiana, India.
8. Santokh Singh; Spiritual Awakening Studies; 2000; Ontario, Canada.
9. The Bible, Mathew 2:1
10. Mcleod WH; Guru Nanak and the Sikh Religion; 1976; Oxford University Press; Delhi, India.
11. Cunningham J.D.; History of the Sikhs; 1997; DK Publishers; New Delhi, India.
12. Dhillon H; The First Sikh Spiritual Master; 2005; Skylight Paths Publishing; Woodstock, Vermont, USA.
13. Singh, Dalip; Shabad Guru Surat Dhun Chela; 2003; B Chattar Singh, Jiwan Singh Press; Amritsar, India.
14. Macauliffe, Max; The Sikh Religion,1963 New Delhi, India

15. Swarn S Bains; Guru Nanak's Divine Teachings; Xlibris Publishers; USA.
16. Isaacson Walter; Einstein: His Life and Universe; 2007; Simon and Schuster; New York.
17. Einstein, Albert and Leopold Infeld; The Evolution of Physics; 1938; Simon and Schuster; New York New York.
18. Bernstein Jeremy; A Theory for Everything; Springer-Verlag; 1996; New York.
19. Aczel Amir; GOD's Equation; Random House; 1999; New York, New York.
20. Hawking, Stephan; A brief history of Time; 1988; Bantam Books; New York.
21. Jammer, Max; Einstein and Religion; Physics and Theology; Princeton University Press; 1995; Princeton, New Jersey.
22. Norton, John D.; Thought Experiments in Einstein's Work; Roman Littlefield; 1991; Savage Mirrorland, USA.
23. Einstein, Albert; Relativity: The Special and General Theory; 1995; Random House Edition; New York, USA.
24. Einstein, Albert; The World as I See It; 1949; New York, USA.
25. Einstein, Albert, Green Brian; The Fabric of the Cosmos: Space, Time and the Texture of Reality; 2004; Knopf, New York, New York.
26. Kaku, Michio; Einstein's Cosmos: How Albert Einstein's vision transformed our understanding of space and time; Atlas Books; New York.
27. Dannen, Gene; Trinity Test, July 16, 1945, Wikipedia, Nov 2014.
28. Goldsmith, Maurice and others; Einstein: The First Hundred Years; Paragon Press; 1980; New York, New York.
29. Powel, Coray; God in the Equation; 2002; Freeport Press; New York, New York.
30. Ryan, Dennis; Einstein and the Humanities; 1987; Greenwood Press; New York, New York.
31. Singh, Simon; Big Bang: The Origin of the Universe; 2004; Harper Collins; New York, New York.

32. Lightman, Alan; Einstein's Dreams; 1993; Pantheon Books; New York, New York.
33. Calaprice, Alice; The New Expanded Quotable Einstein; 2005; Princeton University Press, Princeton, New Jersey, USA.
34. Baierlein, Ralph; Newton to Einstein: The Trail of Light: An Excursion of the Wave-Particle Duality and the Special Theory of Relativity; 2001; Cambridge University Press; New** York, New York.
35. Jeffrey, Grant; The Signature of God; 1996; Frontier Research Publications Inc.; Toronto, Canada.
36. Jeffrey, Grant; The Handwriting of God; 1997; Frontier Research Publications Inc.; Toronto, Canada.
37. R. Edwin, Sherman; Bible Code Bombshell; New Leaf Press; 2005; Arizona, USA.
38. Twyman, James F; The Moses's Code; 2008; Hay House; Carlsbad, California, USA.
39. Muktananda, Swamy; I AM THAT; 1978; SYDA Foundation; New York, USA.
40. Joachim, Harold H.; The Nature of Truth; 1906; Clarendon Press; Oxford, United Kingdom.
41. Bohm, David; Wholeness and the Implicate Order; 1980; Routledge Chapman Hills; New York, New York.
42. Bohm, David; Quantum Theory; 1951; Prentice Hall; New York, New York.
43. Haisch, Bernard; The God Theory; 2009; Redwin Weiser; San Francisco, California.
44. Planck, M.; The Universe in the Light of Modern Physics; 1931; G. Allen & Unwin; New York, New York.
45. Braden, Greg; The God Code; 2004; A House, Inc.; Carlsberg, California, USA.
46. Wilson, Edmund; The Dead Sea Scrolls; 1969; Oxford University Press; New York, New York.
47. Chalmers, David; The Conscious Mind; 1996; Oxford University Press; New York, New York.

48. Powell, Corey; God in the Equation; 2002; Free Press; New York, New York.
49. Bernstein, Aaron; Naturwissenschaftliche Volksbücher Volumes 1 - 12; 1870; Howard Stachel, Germany.
50. Bohr, Niels; "Discussions with Einstein"; Ottostern Recollections; 1982; Paris, France.
51. Glanzberg, Michael; Truth; Stanford Encyclopedia of Philosophy PDF; 2018; Stanford University, California, USA.
52. László, Ervin; Science and the Akashic Field; 2004; Inner Traditions; Rochester, Vermont, USA.

GLOSSARY

1. *Adiyogi* – a perfect yogi who has achieved state of uninterrupted unity consciousness.
2. Allah – is the religious name of God in Islam.
3. *Akasha* – this is the Sanskrit meaning of the sky
4. Akashic field – information contained in the cosmos about the whole history, the past, the present, the future in the cosmic web or the Akashic library.
5. Angad - "ang—part of body"; This *angad* is the name given to the successor of Guru Nanak and he was called Guru Angad Dev.
6. artificial intelligence – is intelligence demonstrated by machines, learning or solving problems like humans.
7. astrophysicist – a specialist who specializes in astronomy using the principles of physics and chemistry to determine the nature of stars and galaxies.
8. *atma* or *atman* – the soul, higher self, the spirit, the inner dweller, light atoms
9. A*um* – Om – the sound of universe
10. atheism – absence of belief in God
11. Bible – the holy book of Christians
12. B*hagvad Gita* – the holy book of Hindus
13. Big Bang – the Big Bang is the hypothetical starting point of our galaxy, a pinpoint of infinite concentrations as an event singularity which happened 13.8 billion years ago.
14. Bi-hemispherical synchronization – is also called hemi-sync which is synchronization of both hemispheres of the brain. In this condition, the electrical activity of the brain is balanced and coherent.
15. *Brahma* – is the Hindu god of creation
16. *choga* – a long outer flowing robe; this was worn by Guru Nanak

17. Cosmological constant – it is a source term and is viewed as the mass of empty space or vacuum energy. Einstein used it as a solution of the gravitational field equation.
18. *daswan dwar* – the 10th gate or the third eye centre, the gateway to enlightenment
19. Dark energy – an unknown form of energy which permeates all of space and causes the expansion of the universe to accelerate. It contributes to 68% of the total energy in the observable universe. It uniformly fills the vacuum space. Dark energy is the intrinsic property of space.
20. Ek – one, the one and only one, the basic number, the creative life principle. Ek was used by Guru Nanak to describe the oneness of God.
21. EPR paradox – Einstein, Podolsky and Rosen thought experiments to testify the phenomena of non-locality (see Wikipedia)
22. Epigenetics – the new science of genetics which involves the influence of environmental factors
23. God equation – is the famous equation of Einstein, $E=mc^2$
24. Gravity – gives way to all things with mass or energy and allows objects to gravitate to one another; like the waves in the ocean are due to gravity from the moon
25. hologram – it is a physical structure that uses laser light diffraction to make a 3D image
26. Holographic universe – researchers believe that our universe is a vast and complex hologram. A map of the oldest light of the universe is reconstructed from the afterglow of the Big Bang by a new set of experimental devices.
27. H*ukam* – the order of the divine will and the universe
28. *Janam Sakhis* – narrative life history of Nanak with biographical details
29. *Janeu* – sacred thread worn by high caste Hindu men
30. *Jap* – recite the name of God
31. *Japji* – this was composed by Guru Nanak and has become the morning prayer of Sikhs

32. *kirat karna* – labour honestly
33. *kirtan* – singing hymns from the holy book
34. Luitpold Gymnasium – this was the high school Einstein attended in Germany
35. *langar* – it is the community kitchen where the rich and the poor and the common man sit together to eat
36. Mardana – constant companion of Guru Nanak. A boyhood Muslim friend who played *rabab*.
37. *mussalman* – a Muslim person
38. metaphysical – is the branch of philosophy that discusses the nature of reality and various relationships, e.g. existence, space, time and other esoteric elements
39. *mantra* – chanting sacred words that may have profound effect on the psyche
40. *moolmantra* – this is the preamble to the sacred opening stanza of *japji*
41. monotheism – belief in the existence of one God
42. *Naam* – the essence of existence, the true name of lord master
43. *Naam Simran* – meditation of the name of God. This is Nanak's use of meditation technique.
44. non-locality – particles, when they are entangled with each other, are not only in one plane or at a distant plane but exist simultaneously in all measured places at the same time.
45. omnioneness – all oneness, totality of oneness
46. omniscience – all knowing
47. omnipresence – all present eternal existence
48. omnipotence – all powerful
49. *Omkar Bani* – it is also called *Ramkali Dakhani* on page 929–938 in SGGS. This was a discourse given by Guru Nanak to the pundit in Omkar temple.
50. Ontology – study of the nature of being, the philosophical discussion on study of being
51. *Oora* – the first letter of *gurmukhi* or Punjabi

52. paganistic – a polytheistic religion called paganism, a form of nature worship and rituals
53. Pantheism – god is the universe itself. This is what Guru Nanak was preaching.
54. Particle – in physics this is a small localized object which has certain properties like volume, density or mass. The size varies from the size of an electron to the size of galaxies.
55. *patti likhi* – writing on a wooden board (*patti*), composition by Guru Nanak at age 7 for his school
56. polytheism – worship of multiple deities with pantheon of gods and goddesses
57. proton – subatomic particles made up of three quarks
58. Pythagorean theorem – equation relating the three sides of a right triangle, named after Greek thinker Pythagoras
59. Quantum – in physics, this is the minimum amount of any physical property involved in an interaction, and quanta is the plural. A photon is a single quantum of light.
60. Quantum mechanics – branch of physics that helps to formulate the laws of nature which describes nature at the atomic and subatomic levels
61. Quantum entanglement – "the observed physical phenomena that occurs when a pair of particles interact or share space in a way that quantum state particles interact… a quantum state of each particle or group cannot be described independently of the state of others even when the particles are separated by a large distance of light years "(see Wikipedia for details.)
62. quarks – these are fundamental particles which combine to form composite particles called hadrons, protons, neutrons
63. Quran – the holy book of Muslims
64. *rabab* – a stringed instrument used by Mardana and is played with a bow
65. *rishi* – man of great learning, living as an ascetic and a life of a hermit
66. *siddha* – a person with spiritual powers

67. *sach* – it literally means truth
68. *Sangat* – a gathering of spiritual people
69. *Sachkhand* – the region or dominion of truth, the residence of lord master
70. *sahaj* – the state of calm where there is inner harmony and equipoise
71. *Siddh Gosht* – this is the discourse and discussions Nanak had with the *Siddha* yogis who questioned him
72. satori – a brief moment of enlightenment or aha
73. Singularity – describes the moment when a civilization changes, like a complete makeover by extremely rapid technological or spiritual profound changes, like a technological singularity or a divine singularity event (see Wikipedia for details)
74. space-time continuum – a mathematical model which fuses the three dimensions of space and one dimension of time into a single four dimensional continuum
75. Soul – the essence of a living being; also called *atman*, self or *jiva*
76. Spirit – it is the vital force within all living things, a non-physical entity also a subtle substance. It's also called the holy spirit.
77. Sri Guru Granth Sahib (SGGS) – the Holy book of the Sikhs; regarded as the living Guru
78. *Shabad* – the word, the true name, the Guru according to Guru Nanak
79. *Surat* – the thinking capacity of the mind
80. *Suras* – verses from the Quran
81. teleportation – there is non-local connectivity and in quantum teleportation of entangled particles, one can create incredible speed of transfer of information at 100 times speed of light
82. *Tera* – *tera* means yours or thou, or the number 13. Guru Nanak used to call *tera tera tera*, which means thou, thou, thou.
83. third eye centre – it is the anatomical equivalent to the pineal gland or the *Daswan Dwar* or the 10th gate
84. *Trimurti* – the three gods coexisting as Brahma, Vishnu, and Shiva in Hinduism

85. trinity – god existing in three persons; the Father, the Son, the Holy Spirit
86. U*dasi*s – itineraries or the missionary travels of Nanak
87. Unified field – it is the fountainhead of all the laws of nature, all the fundamental forces governing, sustaining, and maintaining the whole universe
88. Vacuum energy – it is also called vacuum space. It is devoid of matter in the background energy that exists in space throughout the entire universe. Vacuum has a vastly complex structure and it has similar properties like spin or charge and polarity. Most of these properties cancel out leaving the vacuum—empty, with vacuum fluctuations. In vacuum, particle – antiparticle annihilate each other and disappear.
89. V*eda*s – the ancient and foundational scriptures or Hinduism
90. W*aheguru* – name of God in Sikhism. Also, name of almighty for recitation to keep the spiritual current flowing.
91. the word – proceeds from the point of absolute origin of the creation
92. *yogi* – spiritual practitioner
93. Zero point energy – this is the lowest possible energy; relates to the quantum vacuum energy which is infinite. The vacuum energy is a special case of zero point energy (for details see Wikipedia).

INDEX

Aham Bharmasmi, xiii
Albert Einstein, 12, 15, 25, 31, 44, 60, 67, 76,
atma, 48, 52, 53, 67, 70, 73, 76, 82
Bible code, xiii, 52, 55
Brahama's code, 9
Code, xii, xiii, xiv, xv, 12, 23, 27, 44,-47, 50, 52, 54-62, 64-67, 76, 82, 84-86
dharma, xi, 77
Ek, xiii, xv, 1-2, 36-38, 42, 46, 48-50, 64-67, 73, 76-82, 85-86, 92
Ek Om Kar, xiii, xv, 1-2, 36-38, 46, 48-50, 54-64, 67, 73, 76, 78, 80, 82, 85-86
God, 5, 7-9, 12, 17, 19, 25-39, 43, 45, 50, 54, 56-58, 61, 64, 66-67, 69, 73, 76-77, 80
Guru Nanak, xi, 1-2, 10, 25, 31-32, 35-38, 44, 48, 50-51, 55-57, 60, 62, 65, 67-68, 73, 76-77
Hukam, 46
karma, xi, 34, 77

maya, xi, 80
Naam Simran, xii, 48, 63, 68
Nankana Sahib, 2
neutron, 18, 65
Oneness, 50, 55, 59, 62, 66, 76, 84-85
photoelectric effect, 16-17
Quantum, xii, xiv-xv, 12, 16-17, 23-24, 30-31, 41-46, 48, 50, 57, 60, 63, 66-67, 71-73, 77, 79, 83-84, 94
relativity, xii, 15-16, 24, 42, 72, 83
shabad, 2, 37, 48, 63, 68, 76-77, 85
Sikhs, xiii, 2, 9, 37, 54
Sri Guru Granth Sahib, xiii, 2-3, 36
Supramental Beings, 62, 67, 69
totality, 24-25, 28, 42, 54, 58, 65, 67-68, 93
Unified Field, xii-xv, 16-17, 19, 23, 29, 32, 35, 41, 43, 45-48, 50, 57-60, 62-64, 67, 70-72, 76, 78-80, 83-84, 96
Unified field theory, xii, 17, 67, 83

www.ingramcontent.com/pod-product-compliance
Lightning Source LLC
Chambersburg PA
CBHW021442210526
45463CB00002B/618